Basic Physics

Wiley Self-Teaching Guides teach practical skills from accounting to astronomy, management to mathematics. Look for them at your local bookstore.

Other Science and Math Wiley Self-Teaching Guides:

Science

Astronomy: A Self-Teaching Guide, by Dinah L. Moche

Biology: A Self-Teaching Guide, by Steven D. Garber

Chemistry: Concepts and Problems, A Self-Teaching Guide, Second Edition, by Clifford C. Houk and Richard Post

Math

All the Math You'll Ever Need: A Self-Teaching Guide, by Steve Slavin

Geometry and Trigonometry for Calculus, by Peter H. Selby

Practical Algebra: A Self-Teaching Guide, Second Edition, by Peter H. Selby and Steve Slavin

Quick Algebra Review: A Self-Teaching Guide, by Peter H. Selby and Steve Slavin

Quick Arithmetic: A Self-Teaching Guide, by Robert A. Carman and Marilyn J. Carman

Quick Business Math: A Self-Teaching Guide, by Steve Slavin

Quick Calculus: A Self-Teaching Guide, Second Edition, by Daniel Kleppner and Norman Ramsey

Statistics: A Self-Teaching Guide, by Donald Koosis

Basic Physics

A Self-Teaching Guide

Second Edition

Karl F. Kuhn

John Wiley & Sons, Inc.

New York • Chichester • Brisbane • Toronto • Singapore

Library of Congress Cataloging-in-Publication Data:
Kuhn, Karl F.
    Basic physics / Karl F. Kuhn. — 2nd ed.
      p.   cm. — (A self-teaching guide)
    Includes index.
    ISBN 0-471-13447-3 (pbk. : alk. paper)
    1. Physics. I. Title. II. Series.
  QC23.K74   1996
  530–dc20   95-38583

Printed in the United States of America
20 19 18 17 16 15

To Christ J. Kuhn, my paternal grandfather.
"Dodo" taught me what a friend is.
If I can do the same for my grandchildren,
my life will have been a success.

Acknowledgments

First, I am grateful to the many students who used the first edition of *Basic Physics* and made suggestions that helped form this edition. A big thank you to Karyn West, who keyed the entire manuscript into her trusty Macintosh, and to Kimberly Carter, who served as my personal editor and grammar checker. I would love them even if they were not my daughters.

Thanks for the tremendous help and understanding from Judith McCarthy, John Cook, Lynne Frost, and the rest of the team at John Wiley and Sons, Inc.

I give my gratitude and love to those who make the work worthwhile: my wife Sharon, my daughters Karyn and Kim, my sons Deuce, Keith, and Kevin, and my grandchildren Amanda, Jim, Ty, Parker, Patrick, Adam, and two yet to be named. They give more than they take.

Contents

To the Reader   *xi*

1   Force and Motion   1

2   Newton's Laws of Motion   13

3   The Conservation of Momentum and Energy   24

4   Gravity   41

5   Atoms and Molecules   51

6   Solids   64

7   Liquids and Gases   71

8   Temperature and Heat   84

9   Change of State and Transfer of Heat   94

10 Wave Motion   103

11 Sound   113

12 Diffraction, Interference, and Music   123

13 Static Electricity   138

14 Electrical Current   149

15 Magnetism and Magnetic Effects of Currents   164

16 Electrical Induction   176

17 Electromagnetic Waves   189

18   Light: Wave or Particle?   197

19   The Quantum Nature of Light   212

20   Reflection, Refraction, and Dispersion   231

21   Lenses and Instruments   249

22   Light as a Wave   270

23   Color   282

Appendix I   Scientific Notation: Powers of Ten   289

Appendix II   The Metric System   293

Index   295

To the Reader

"If it ain't broke, don't fix it." This maxim applies to books as well as to many other endeavors, and it explains why such a long time passed between the first and second editions of this book. Elementary physics does not change much in a few decades, and therefore this edition contains very little new subject matter.

Why a new edition, then? First, some new examples have been added, either because of change in fashions or because of new applications of basic physics. An example of the latter is the now-common use of fiber optics, an application that was just coming on the scene when the first edition was printed. Second, some explanations have been rewritten to improve clarity. Finally, the "look" of the book has been updated to make it more attractive to today's readers.

The fundamental purpose of the book has not changed, however, and if you are considering a study of this book you probably fit into one of the following categories:

- You are enrolled in an introductory physics course and wish to have a study guide to accompany your textbook. This book follows the traditional order of topics that is used in the most popular classroom texts.

- You are taking a course that uses a nonmathematical, conceptual book as its primary text, and your instructor has adopted this book as a supplement in order to include more problem-solving in the course. The traditional order of topics in this book is also used in the most prevalent conceptual physics texts.

- You are enrolled in a course that uses this book as its primary text.

- You are not taking a formal physics course, but you wish to learn, or review, some physics—perhaps to enable you to pass a test that requires a knowledge of physics.

In any case, *Basic Physics* can help you. It is a complete, self-contained physics book with a programmed format. The chapters are divided into short steps called frames. Each frame presents some new material, and then asks you questions to test your comprehension. By faithfully answering the questions (preferably by actually writing the answers in the spaces provided or on a separate sheet) before you check the answers I supply, you will be able to check your progress. Learning theory tells us that by writing your answers, you understand and retain the material for a longer time. In fact, I suggest that you cover the furnished answers with a card or your hand until you have completed your own. If your answer does not agree with the one I provide, be sure you understand the discrepancy before proceeding to the next frame. To check yourself further, a Self-Test, with answers, is included at the end of each chapter.

Since physics builds from one principle to another, many chapters require an understanding of previous chapters. For this reason, the prerequisites for each chapter are listed on the first page of the chapter. Some frames within chapters, however, may be skipped without disrupting the logic of the development. Such frames, which are often mathematical treatments of the subject at hand, are labeled as "optional." Some optional frames have a prerequisite of a prior optional frame; such prerequisites are listed at the beginning of the frame.

I suggest that you complete the entire Self-Test at the end of a chapter before checking your answers. In this way, your test will be more similar to classroom testing. Each answer in the Self-Test section includes a reference to the frame(s) to which you should return for help on missed items.

Have fun!

1 Force and Motion

Physics deals with quantities that can be measured. Thus, you won't find concepts such as honesty, love, and courage as primary topics of discussion in a physics book. As you proceed through your study of physics, you will find that every one of the measurable quantities that is discussed can be specified in terms of only four basic dimensions: mass, length, time, and electric charge. In this chapter, we will begin a study of the first three of these.

OBJECTIVES

After completing this chapter, you will be able to

- define speed;
- calculate speed, distance, or time, given the other two values;
- calculate acceleration, given initial and final velocities and the elapsed time;
- state the value of the acceleration of gravity in both the metric and the British systems of units;
- given an initial velocity, calculate the distance an object will fall in a given time;
- differentiate between speed and velocity;
- distinguish between scalar and vector quantities, giving an example of each;
- use the graphical method to add vectors;
- identify and classify (as to acceleration and velocity) each of the two components of projectile motion.

1 SPEED

Speed is defined as the distance traveled divided by the time of travel. Thus, if you travel 12 miles in 3 hours, the average speed is 4 miles per hour* (mi/hr or mph).

---

*In using miles per hour, you are using the British system of units, the one you may be most familiar with. It includes such units as inches, feet, and pounds. I am using this system to introduce some concepts, but most of our work will use the metric system.

(a) If an elephant runs for half an hour at a speed of 6 miles per hour, what distance does it cover? _____

(b) What is the average speed of a vehicle that covers 250 miles in 2.5 hours?

_____

(c) Using the symbols s, d, and t (for speed, distance, and time), write a formula for speed. _____

Answers: (a) 3 miles;   (b) 100 mph;   (c) $s = d/t$

In practice, we usually use the symbol v to represent speed (because of the similarity between speed and velocity, a concept that will be discussed later). The defining equation for speed is:

$$v = \frac{d}{t}$$

Here v is the speed and d is the distance from the object's starting position to its position after time t has elapsed.

Physicists use the SI system (Système International), which uses meters as its basic unit of length. If you walk with a speed of 2 meters per second (m/s), how far do you walk in 12 seconds? _____

Answer: 24 meters

ACCELERATION

As a car gains speed, we say that it accelerates. The change in speed divided by the time it takes to make the change is called the **acceleration**. In equation form:

$$a = \frac{v_1 - v_0}{t}$$

Here v_0 represents the speed at the start and v_1 is the speed after time t has passed.

Notice that although in everyday language the word "acceleration" is used only to refer to a gain in speed, if v_0 is larger than v_1, the acceleration is negative and is actually a deceleration. In physics we use the word "acceleration" to include this negative case.

(a) What part of the equation represents the change in speed? _____

(b) If a car takes 5 seconds to increase its speed from 30 miles/hour to 50 miles/hour, what is its acceleration? _____ miles per hour per second.

Answers: (a) $v_1 - v_0$;   (b) 4

4

The unit of acceleration in the problem you just solved is "miles per hour per second," or "mi/hr/s." If the speed is measured in meters per second and its change is measured over a few seconds, a natural unit for acceleration is "m/s/s," which is sometimes called meters per second squared and written "m/s²."

Suppose a car starts from rest and accelerates at 2 m/s² for 3 seconds. This means that the car gains _____ of speed during each _____.

Answer: 2 m/s...second

Suppose that, starting from rest, your new truck accelerates at the rate of 5 mi/hr/s and it continues this acceleration for 8 seconds. What will be its speed at the end of the 8 seconds? _____ (If you don't know how to find the answer, read the rest of this frame; otherwise, just calculate the answer.)

To solve this problem, first write the acceleration equation:

$$a = \frac{v_1 - v_0}{t}$$

In this case, v_0 is zero since the car started from rest. Now solve the equation for v_1, the unknown speed:

$$v_1 = at$$

Substitute in the acceleration and time, and solve for v_1.

Answer: 40 mi/hr

THE ACCELERATION OF GRAVITY

The acceleration of an object falling freely near the surface of the earth is 9.8 m/s², which is 32 ft/s² in the British system. An object dropped from a very high platform achieves a speed of 32 ft/s by the end of the first second and 64 ft/s by the end of the next second.

Suppose an object starts from rest and falls freely. After 1 second its speed is 9.8 m/s.

(a) What is its speed in m/s after 2 seconds? _____

(b) What is its speed after 3 seconds? _____

Answers: (a) 19.6 m/s; (b) 29.4 m/s

7

The figure on the next page shows a ball being dropped from a building, and it indicates the speed of the ball at various times while it falls.

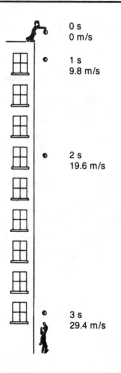

0 s
0 m/s

1 s
9.8 m/s

2 s
19.6 m/s

3 s
29.4 m/s

Now suppose that a ball is thrown upward from the base of the building at a starting speed of 29.4 m/s. On the way up its acceleration (or deceleration, if you wish) would be 9.8 m/s², just as it was on the way down, except that now the acceleration is downwardly directed while the speed is upward. Because the acceleration is the same whether the object is moving up or down, the figure could just as well represent the person at the bottom throwing the ball upward.

In this latter case, what would be the speed of the ball when it reaches the top? _____

Answer: 0 m/s

---

8 ACCELERATION EQUATIONS

The equation relating acceleration, the distance traveled, and the time of fall is:

$$d = v_0 t + \frac{1}{2} a t^2$$

where v_0 is the original speed and t represents time elapsed. Thus, in the figure in frame 7, when the ball has fallen 1 second (and its speed is 9.8 m/s), it has fallen 4.9 meters $\left(d = \frac{1}{2} \cdot 9.8 \cdot 1^2 \text{, since } v_0 \text{ was } 0 \right)$.

After 3 seconds of fall how far has it fallen? _____

Answer: 44.1m $\left(d = \dfrac{1}{2}at^2 = \dfrac{1}{2}(9.8)(3)^2 = 44.1\right)$

9   Although its speed after 1 second is 9.8 m/s, the ball fell only 4.9 meters during the second. This is because it started from rest—at zero speed—so its average speed during the first second is 4.9 m/s, or half of 9.8. Consider the first 3 seconds of fall.

Initial speed = zero

(a) Final speed: $v_1 = at =$ _____

(b) Average speed = _____

Answers: (a) 29.4 m/s;   (b) 14.7 m/s

(Part (b) was calculated by considering that since the ball fell for 3 seconds at an average speed of 14.7 m/s, its distance traveled is "14.7 m/s times 3 seconds," which equals 44.1. This corresponds to what you obtained by use of the equation $d = v_0 t + \frac{1}{2}at^2$.)

10   # VELOCITY—A VECTOR QUALITY

Often speed does not tell us all we want to know about a motion. If you were told that while you slept your sheepdog left you at a speed of 5 mi/hr, and that he traveled in a straight line, you would not have enough information to go looking for him. You would also need to know which direction he went. Although the term **speed** tells us how fast something is moving, the term **velocity** specifies both speed and direction. Identify the two descriptions below as speed or velocity.

(a) 20 meters/second _____

(b) 20 meters/second northeast _____

(c) How does velocity differ from speed? _____

Answers: (a) speed;   (b) velocity;   (c) Velocity includes a direction.

11

The defining equation for velocity is:

$$v = \frac{d}{t}$$

Notice that velocity is defined in the same way as speed, except that in this case we print some symbols in boldface. The quantities in boldface are **vectors**. A vector is a quantity that has direction as well as magnitude (size). The symbol **d**, called the object's **displacement**, is defined as the *straight-line* distance from the object's starting position to its final position, including the direction from start to finish. Thus, an object's displacement might be stated as 4.2 meters toward the east.

(a) Which quantities in the equation above are vectors? _____

_____

(b) Which are not vectors? _____

Answers: (a) velocity and displacement (**v** and **d**);   (b) time (*t*)

12

We cannot assign a special direction to time, so it is not a vector. All quantities that are not vectors are called **scalars**. Identify each of the following quantities as a vector or a scalar.

(a) velocity _____

(b) speed _____

(c) time _____

(d) displacement _____

Answers: (a) vector;   (b) scalar;   (c) scalar;   (d) vector

13

In the next chapter, you will see that acceleration is also a vector, since we must take into account its direction. (We have already seen that a negative acceleration is different from a positive one.)

To see the importance of vectors in a simple situation, consider the following example: Suppose a jogger leaves her house and jogs north 300 feet to the streetlamp on the corner. She turns west, jogs another 400 ft, and gets to the oak tree 2 minutes after leaving home. We will calculate the jogger's average velocity for the 2 minutes. (See diagram on next page.)

(a) Which of the three distances given in the figure is the displacement **d**?

_____

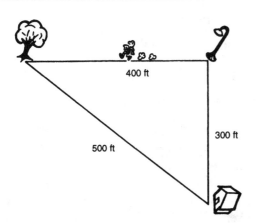

(b) What is the jogger's average velocity?

_____

_____

(c) What is the total distance (not displacement) traveled by the jogger? _____

(d) What is the average speed (not velocity) during the 2 minutes? _____

Answers: (a) 500 feet; (b) 250 ft/min in a direction between north and west; (c) 700 feet; (d) 350 ft/min

This example shows that velocity and speed are definitely different things. It may seem to you that speed is the more useful of the two, but the following examples should show the advantage of the concept of velocity.

14 A VECTOR APPLICATION—PROJECTILE MOTION

Imagine a moving hot-air balloon that drops a sandbag. Since the balloon is moving along with the air, the effect of air resistance can be neglected as the sandbag falls. An interesting thing occurs: as the sandbag falls, it continues moving forward at the speed that the balloon had when the sandbag was dropped. The sandbag does fall, of course, but its downward speed is independent of its forward speed. The sandbag continues forward as if it had not been dropped, and it falls downward with an ever-increasing speed just as if it were not moving forward.

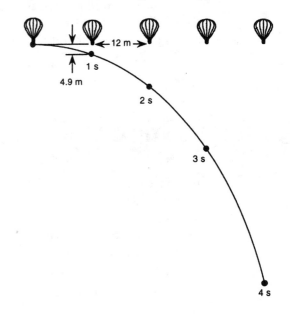

The figure assumes that the balloon is traveling at 12 m/s, and it shows the sandbag at intervals of 1 second beginning when it is dropped. In each second the sandbag moves forward 12 meters. Now note its downward motion. At the end of the first second after being released, its *downward* speed is 9.8 m/s.

(a) At the end of 2 seconds, what is the sandbag's forward speed? _____

(b) At the end of 2 seconds, what is its downward speed? _____

(c) At the end of 3 seconds, what is its forward speed? _____

(d) At the end of 3 seconds, what is its downward speed? _____

Answers: (a) 12 m/s; (b) 19.6 m/s; (c) 12 m/s; (d) 29.4 m/s

In this example, we have seen that it is sometimes advantageous to consider motion as having two components, each perpendicular to the other. Although the actual velocity of the sandbag is neither straight down nor horizontal, we can consider its motion to be a combination of a downward motion and a horizontal motion. When we do this, each of the separate motions is a simple one. We will return to projectile motion later in this chapter and again in the next.

15 ADDING FORCES—ANOTHER VECTOR APPLICATION

Suppose two children are having a tug-of-war over their favorite toy—a red wagon. One child is using a force of 9 pounds to try to pull the wagon toward her house. The other child also pulls with a force of 9 pounds, but he is pulling in the opposite direction. This means that one force balances the other. They are equal in magnitude (size) but opposite in direction, so they *add* to zero. Since we need to consider the direction of forces in order to add them, we see that force is a vector quantity.

Often we are interested in the resultant, or **net**, force on an object. Suppose that the first child pulls one way with a force of 11 pounds and the other pulls in the opposite direction with a force of 10 pounds.

What is the net force on the toy in this case? _____

Answer: (a) 1 pound (in the direction of the first child)

16 MATHEMATICAL METHODS OF VECTOR ADDITION (OPTIONAL)

Since force is a vector quantity, we must take into account their directions in order to add two or more forces. Addition of vectors is most easily accomplished by using scale drawings—the "graphical" method of vector addition. Suppose two cars push on the huge pie in part 1 of the figure at the top of page 9. Part 2 uses arrows to represent the forces the cars exert. Since the compact car exerts 160 pounds of force and the large one 80, the arrow representing the small car's force is twice as long as the large one's (4 cm and 2 cm, respectively). In part 3,

(1)

(2)   80 lb (2 cm)

(3)

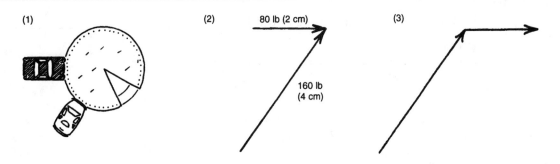

160 lb
(4 cm)

the arrows are shown hooked end-to-end, but they each keep the same length and are pointed in the same direction as the forces exerted.

Now *you* draw a line from the tail of the first arrow (the long one) to the head of the second one. This vector will represent the net force, so place an arrowhead at its upper end. Go ahead and draw it now, then measure it. What is its length?

_____

Answer:

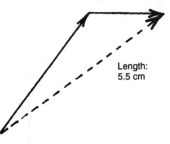

Length:
5.5 cm

17

Since we used 1 cm to represent 40 pounds, your resultant vector represents a force of 220 lb (5.5 times 40) pointing in a direction between the two cars' directions of force. Note that it is, as should be expected, a little closer to the direction the small car is pushing. (This is because that car is pushing much harder and thus affects the resultant direction more than does the larger car.)

(a) Using the same scale (where 1 cm represents 40 lb), use the space below to draw the vector diagram if the compact car pushes with a force of 140 lb and the large one with 200 lb.

(b) What is the resultant force in this case? _____

Answers:
(a)

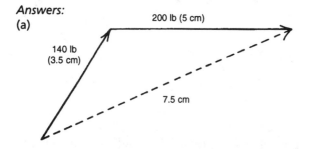

200 lb (5 cm)

140 lb
(3.5 cm)

7.5 cm

(b) 300 lb (If your resultant vector measures between 7.0 and 8.0 cm, that is OK. You should get between 285 and 320 lb.)

18 Vector addition also is used with velocities. Consider an airplane flying in still air at a velocity of 150 miles per hour northward. In part 1 of the figure below, a 3-cm arrow represents this velocity. Now a wind starts blowing toward the southeast at a speed of 50 mi/hr—quite a wind! The second part of the figure shows this 50-mi/hr wind as a 1-cm arrow. Just as was done in the addition of forces, one arrow is moved so that its tail falls on the head of the other, shown in part 3.

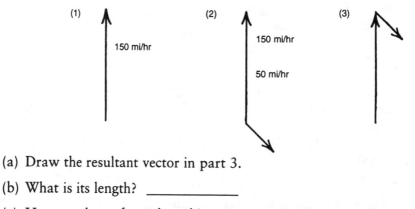

(1)

150 mi/hr

(2)

150 mi/hr

50 mi/hr

(3)

(a) Draw the resultant vector in part 3.

(b) What is its length? _____

(c) How much net force does this represent? _____

Answers: (a) It is drawn from the tail of the first to the head of the second. (b) 2.4 cm; (c) 120 mi/hr (because each centimeter represents 50 mi/hr) (Note that the plane is blown off course as well as being slowed down.)

19 Let us return once again to projectile motion and the example of the sandbag being dropped from the hot-air balloon. Suppose the balloon is high enough that the sandbag hits the ground 2 seconds after it is released. At that time the forward speed of the sandbag is still 12 m/s. Its downward speed is now 19.6 m/s. On a separate sheet of paper, use the graphical method to combine both vectors, and determine the velocity of the sandbag at impact.

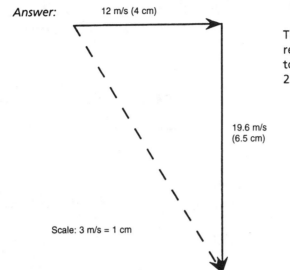

Answer:

12 m/s (4 cm)

19.6 m/s
(6.5 cm)

Scale: 3 m/s = 1 cm

This drawing uses a scale of 1 cm to represent 3 m/s. The arrow that shows the total is 7.6 cm long, so the total velocity is 23 m/s.

SELF-TEST

The questions below will test your understanding of this chapter. Use a separate sheet of paper for your diagrams or calculations. Compare your answers with the answers provided following the test.

1. Define speed. _____

2. A train goes 300 miles at an average speed of 45 miles per hour. How long does it take to go the distance? _____

3. An object is traveling with a speed of 2 m/s. Three seconds later its speed is 14 m/s. How much is its acceleration during this time? _____

4. The acceleration of gravity in British units is _____. The acceleration of gravity in metric units is _____

5. How far does the object in question 3 travel during the 3 seconds of its accelerated motion? _____

6. An object falling from rest reaches a speed of _____ ft/s (or _____ m/s) after falling 5 seconds. (Ignore air resistance.) At this time it has fallen _____ ft (or _____ m).

7. Distinguish between speed and velocity. _____

_____

8. Suppose you are a passenger in a car moving at 55 miles per hour. You lift a pencil to the ceiling of the car and drop it. Just before it hits the seat, how fast is it traveling forward? _____ What is happening to its downward speed at this time? _____

_____

9. One boy pulls on his bicycle with a force of 15 pounds. His (former) friend pulls on it with a force of 18 pounds in the opposite direction. What is the net force on the bike? _____

10. *Optional.* A force of 3 pounds is exerted toward the east on an object. At the same time a 4-pound force toward the north is exerted on it. What is the net force on the object? _____ Describe the direction of this force.

_____

ANSWERS

If your answers do not agree with those given below, review the sections indicated in parentheses before you go on to the next chapter.

1. Distance traveled divided by the time used to cover that distance. (frame 1)

2. $6\frac{2}{3}$ hours

    Solution:

    $$s = \frac{d}{t}$$
    $$t = \frac{d}{s} = \frac{300 \text{ miles}}{45 \text{ mi/hr}} \quad \text{(frames 1, 2)}$$

3. 4 m/s/s, or 4 m/s² (frames 3–5)

4. 32 ft/s²; 9.8 m/s² (frame 6)

5. 24 meters (frame 8)

6. 160 (49); 400 (122.5) (frames 6–9)

7. Velocity includes the direction of an object as well as the object's speed (which does not include direction). (frames 10–13)

8. Its forward speed is still 55 miles/hour; Its downward speed is increasing at the acceleration of gravity. (frame 14)

9. 3 pounds toward the former friend (frame 15)

10. 5 pounds; The direction is east of north, or if you have a protractor you can measure it to be 37° east of north. (frame 16)

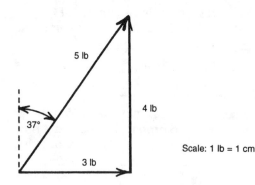

2 Newton's Laws of Motion

Prerequisite: Chapter 1

In the previous chapter you studied the definitions that allow us to describe motions of various types. In this chapter you will use those rules along with the powerful insight into motion that is provided by Isaac Newton's three laws of motion.

OBJECTIVES

After completing this chapter, you will be able to

- state and give an example of the application of each of Newton's laws of motion;
- distinguish between mass and weight;
- given two of the three quantities—force, mass, and acceleration—calculate the third;
- calculate the weight of an object of known mass;
- use Newton's second law to explain why objects of different mass fall with the same acceleration;
- specify the cause of terminal speed;
- give an example of an object moving with constant speed while at the same time accelerating;
- relate Newton's second law to circular motion;
- identify the force-pairs of Newton's third law for cases of both accelerated and nonaccelerated motion.

1 MASS AND INERTIA: NEWTON'S FIRST LAW

> An object at rest tends to stay at rest and an object in motion
> tends to stay in motion in a straight line at a constant speed.

The above is a statement of **Newton's first law,** and it expresses what we mean when we say that an object has **inertia.** Inertia is that property of matter which causes matter to resist change in motion. Mass (one of the four basic measurements) is a measure of the amount of inertia an object possesses.

(a) Which has more inertia, an object of mass 6 kilograms (kg) or one with a mass of 3 kg? _____

(b) Which has more of a tendency to remain at the speed at which it is going?

_____

(c) Which is more likely to hurt, being hit by a bowling ball or by a marble if both have the same speed? _____

(d) Why does it hurt a barefooted person more to kick a bowling ball than to kick a marble? _____

Answers: (a) 6;   (b) 6 kg;   (c) bowling ball;   (d) The bowling ball has more mass—or more inertia, which means the same thing.

If you said, in your answers to parts (c) and (d), that you would be hurt more by one object than the other because one object has more weight than the other, you are not quite correct, as will become clear after more discussion of weight and gravity.

2   NEWTON'S SECOND LAW

If one object has more inertia than another, more force is required to give that object the same acceleration as the first. Suppose that you want an object to accelerate at some specific rate, such as 5 m/s². The amount of force you must apply depends upon the mass of the object. In fact, if you have an object with twice the mass of another one, you must use double the force to get the same acceleration for the first object. And for three times the mass, you need three times the force. You can see that, for a given mass, the acceleration produced is proportional to the force used.

The relationship between force, mass, and acceleration is summed up in a single equation, **Newton's second law:**

$$\mathbf{F} = m\mathbf{a}$$

where **F** is force, m is mass, and **a** is acceleration. Note that the symbols for force and acceleration are in bold type, indicating that these are vectors.

(a) Is mass a vector or a scalar? _____

(b) Which would require a greater force—accelerating a 2-kilogram mass at 2 m/s² or a 4-kilogram mass at 5 m/s²? _____

Answers: (a) scalar (It cannot be assigned a direction.);   (b) 4-kilogram mass at 5 m/s²

3   ACCELERATION AS A VECTOR

In the discussion of acceleration in Chapter 1, we found that if an object is slowing down, we have to assign its acceleration a negative value. In doing this, we are considering acceleration a vector, for in that case the acceleration was in a different direction from the object's velocity.

Refer again to the figure that accompanies frame 7 of Chapter 1, the figure that shows the ball being dropped from the top of the building. We concluded that if the ball had been thrown from the bottom of the building with a speed of 29.4 m/s, it would reach a speed of zero when it reached the top 3 seconds later. Let's consider the ball 4 seconds after it has been thrown upward. We repeat the equation for acceleration, again indicating vectors with bold type:

$$\mathbf{a} = \frac{\mathbf{v}_1 - \mathbf{v}_0}{t}$$

In this case, $\mathbf{v}_0$ = 29.4 m/s, t = 4 seconds, and $\mathbf{a}$ = –9.8 m/s². The minus sign indicates that the acceleration is downward, in the opposite direction of the velocity. If we substitute these values in the equation in order to calculate the final velocity, $\mathbf{v}_1$, we find that

$$\mathbf{v}_1 = -9.8 \text{ m/s}$$

The negative sign indicates that the velocity at this time is downward. (Recall that $\mathbf{v}_0$ was upward—positive—and acceleration was downward—negative.)

Now let's use another equation from the first chapter, and this time we'll include its vector nature:

$$\mathbf{d} = \mathbf{v}_0 t + \frac{1}{2}\mathbf{a}t^2$$

As before, the ball has been thrown upward at a speed of 29.4 m/s. Using the equation for displacement, calculate the position of the ball 4 seconds after it is released. _____

Answer: **39.2 meters above the ground**

4

Note that you did not have to worry about the fact that the ball went to the top and came back down. If you substituted $\mathbf{v}_0$ = 29.4 m/s, t = 4 s, and $\mathbf{a}$ = –9.8 m/s², you should have obtained the answer given, which is the displacement 4 seconds after the ball is released. Since upward is designated as positive, the fact that the answer is positive indicates that the ball is above the ground. Such is the power of using vector methods in the equation.

The fact that the equation $\mathbf{F} = m\mathbf{a}$ is a vector equation tells us that the directions of the force and the resulting acceleration are exactly the same. In the case of the falling ball, the force (which is the force of gravity) and the acceleration are both downward.

(a) When the ball is thrown upward and is on the way up, what is the direction of its acceleration? _____

(b) When the ball is thrown upward and has passed the top of its path so that it is on the way down, what is the direction of its acceleration? _____

(c) If—in a case entirely different from the dropped ball—the accelerating force is toward the southeast, what is the direction of the resulting acceleration?

_____

Answers: (a) downward;   (b) downward;   (c) southeast

In Chapter 1 we analyzed the motion of an object that was dropped from a hot-air balloon. It was possible to make predictions about such a seemingly complicated motion because the vector nature of Newton's second law allowed us to consider vertical and horizontal motions separately. The force of gravity acts downward, and therefore its effect on the sandbag is limited to an acceleration in the downward direction. There was no force on the sandbag in the horizontal direction. Therefore, the sandbag continued its forward motion with no change.

5 UNITS USED IN NEWTON'S SECOND LAW

If mass is expressed in kilograms and acceleration in m/s^2, the correct unit of force is the **newton** (N). A newton is about ¼ pound, the weight of a stick of margarine. Use Newton's second law to solve these problems.

(a) How much acceleration is produced by a force of 12 newtons exerted on an object of 3 kilograms mass? _____

(b) To produce an acceleration of 4 m/s^2 on a bowling ball with a mass of 6 kg, what force would be required? _____

(c) Suppose one wishes to accelerate a 1200-kg Toyota from 10 m/s (which is about 20 mi/hr) to 30 m/s (about 60 mi/hr) in 8 seconds. What force would be required? _____

Answers: (a) 4 m/s^2;   (b) 24 N;   (c) 3000 N

If you were unable to work problem (c), continue below. If you got the answer, you may skip the next paragraph.

6 To solve problem (c) above, work through these steps:

Step 1. Calculate the acceleration involved in changing the speed from 10 m/s to 30 m/s in 8 seconds. _____

Step 2. Use the formula $F = ma$ to find the force required to accelerate a 1200-kg object at that rate. _____

Answers: (a) 2.5 m/s²; (b) 3000 N

7 MASS AND WEIGHT

Suppose you observe a 1-kg object accelerating at the rate of 9.8 m/s². You know that a force of 9.8 N must be accelerating the object. This acceleration of 9.8 m/s² is the acceleration that occurs when an object is released near the surface of the earth—the acceleration of gravity mentioned in the previous chapter. Thus, the force of gravity on a 1-kg object is 9.8 N. We say that a 1-kg object weighs 9.8 newtons.

The foregoing allows us to state a general method for finding the weight of an object whose mass is known. In your imagination, drop an object of mass 10 kg.

(a) How much will its acceleration be? _____

(b) How much is the force that causes the acceleration? _____

(c) How much is the weight of the object? _____

Answers: (a) 9.8 m/s²; (b) 98 N; (c) 98 N

8 We have seen a definite distinction between weight and mass. Weight is a force that is due to the earth's pull of gravity. Mass is a measure of the inertia possessed by an object. Mass is the more fundamental quantity because the amount of inertia of an object does not depend upon its location relative to the earth, but the weight of an object does—an object weighs less as it gets farther and farther from the earth.

(a) Which system, metric or British, commonly uses mass rather than weight?

_____

(b) Which is more stable, the mass or the weight of an object? _____

Answers: (a) metric; (b) mass

If two forces are being exerted on an object at the same time, we must determine the net force, or the resultant force, and use this value in the equation. Frames 15 through 18 of Chapter 1 discussed how vectors can be added to determine their resultants.

9 GRAVITY AGAIN

Consider a 1-kg grapefruit and a 3-kg pumpkin. Now suppose that we apply forces to each object. We exert three times as much force on the pumpkin as on the grapefruit, perhaps 4 newtons on the grapefruit and 12 newtons on the pumpkin. Now apply Newton's second law in each case to calculate the acceleration of each object.

Grapefruit: **a** = _____   Pumpkin: **a** = _____

Answers: 4 m/s²; 4 m/s²

The importance of this simple example lies in the fact that in nature there is a force that depends directly on the mass of the object on which the force is exerted. The force of gravity on an object is directly proportional to the mass of the object. Thus, the weight of the pumpkin is three times as much as the weight of the grapefruit. This means that if we let this force—gravity—act on each of the objects, both objects will have the same acceleration.

Indeed, the pumpkin has three times the mass, but it also has three times the weight. Thus, the equation of Newton's second law yields the same value for the acceleration of any earthly object that falls under the influence of gravity, and Newton's second law explains why the acceleration of gravity is the same for all objects, no matter what their weight. This acceleration is, of course, the acceleration of gravity discussed in Chapter 1.

10

The accleration of gravity is given a special symbol, **g**.

$$\mathbf{g} = 32 \text{ ft/s}^2 = 9.8 \text{ m/s}^2$$

The direction of **g** is downward, of course, so that if upward has been designated positive in using an equation, **g** will be negative.

Objects on the moon weigh less than they do on earth. If we drop an object on the moon, the force of gravity on it will be less, and according to Newton's second law it will have less acceleration. In fact, the acceleration of gravity on the moon is only about ⅙ of the acceleration of gravity on earth.

In considering an object's weight on earth and on the moon, we considered differences in the forces of gravity at the two locations, but we did not consider the differences in the object's mass at those locations. Why? _____

_____

Answer: An object's mass is the same on earth and on the moon.

11 TERMINAL SPEED

In actual practice, falling objects do not continue to accelerate at 9.8 m/s² as they fall. Their failure to continue gaining speed at the same rate is caused by the upward force of air resistance. The acceleration of the object depends upon the net force acting on the object.

If you knew the numerical values needed, how would you calculate the net force exerted on the ball shown in the figure?

Answer: Subtract the air resistance from the object's weight.

12

As downward speed increases, so does air resistance. Thus, there comes a time when air resistance equals the weight of the object. At this speed the object no longer accelerates, but instead it continues at the same speed. (This follows from **F** = *m***a**, since if the net force is zero, the acceleration must also be zero.) The final speed attained is called the object's **terminal speed**.

Consider a falling rock at a time when it is traveling at just less than terminal speed. Will its acceleration be less than, equal to, or more than 9.8 m/s² at this time? _____

Answer: less (When the object finally reaches terminal speed, the air resistance is equal to the rock's weight. At slightly less than terminal speed, air resistance may be considerable, but it is less than the rock's weight. The net force on the rock is its weight minus air resistance. Since air resistance is less than the rock's weight, the acceleration must be less than the acceleration of gravity.)

13 CIRCULAR MOTION

When an object moves in a circle, there is an unbalanced force on it that acts toward the center of the circle. (In the case of a rock whirling on the end of a string, the string provides the force toward the center.) This force is called the **centripetal force**. According to Newton's second law, what happens if an object has an unbalanced force on it? _____

Answer: The object accelerates (or it changes velocity).

14

I hope that you see an apparent conflict between your answer above and your experience. You know that it is possible for an object to move around in a circle at a constant speed. How can there be an acceleration in this case? Remember that acceleration was defined as the change in velocity divided by the change in time, and remember that velocity includes direction. What is happening in the case of an object moving in a circle is that the object's direction is changing even if its speed is not.

Suppose the string in the figure below breaks as you whirl the rock.

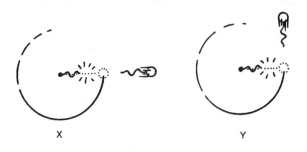

X   Y

(a) Will the rock fly off as shown in X or in Y? _____

(b) Use Newton's laws to explain your answer. _____

_____

Answers: (a) Y; (b) The rock will continue in the same direction if no external force is exerted on it.

15 NEWTON'S THIRD LAW (THE LAST!)

Imagine a 2-pound banana resting on a plate. (That is a big banana!) The banana pushes downward on the plate with a force of 2 pounds. Newton's third law tells us that the plate must therefore push upward on the banana with an equal force. The two forces form the force-pair of Newton's third law, which can be stated as follows:

When object X exerts a force on object Y,
then object Y exerts an equal and opposite force on X.

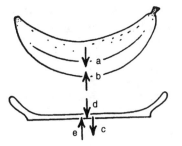

There are two forces on the banana—the downward force, a, of gravity on the banana (which is the banana's weight) and the upward force, b, of the plate on the banana. There are three forces exerted on the plate: the force of gravity on the plate (indicated as c), the downward force, d, exerted by the banana on the plate, and the force e exerted upward on the plate by the table or floor.

Which two forces of the five indicated in the figure form the action–reaction pair of Newton's third law? _____

Answer: b and d

16 Although forces a and b in the preceding example are equal and are in opposite directions, they are not an action–reaction pair of Newton's third law. Newton's law involves (1) the force one object exerts on a second and (2) the force the second object exerts on the first. The tree limb in this figure exerts a 1-newton upward force on the apple (which weighs 1 newton).

What is the reaction force to the force exerted by the limb? To answer this, state the magnitude and direction of the force, then name the object that exerts the force and the object on which the force is exerted.

Answer: The apple exerts a 1-newton downward force on the limb.

17 # NEWTON'S THIRD LAW DURING ACCELERATION

The third law holds that for *every* force exerted by X on Y, there is an equal and opposite force exerted by Y on X. If this is true, there must be a force equal and opposite to the apple's weight. To see what this opposite force is, we must ask what object exerts the gravitational force on the apple. The earth does. Thus, the reaction force is the force of gravity exerted by the apple on the earth. Think of gravity as being a stretched spring between the apple and the earth. There is a force downward on the apple and upward on the earth.

But what if the apple falls? While it is falling, Newton's third law still applies. The only force then acting on the apple is the force of gravity. Since it is not balanced by another force, the apple accelerates—according to Newton's second law. The apple likewise continues to exert a force upward on the earth. Why, then, don't we see the earth rise to meet the falling apple? (Hint: Apply Newton's second law to the earth.) _____

Answer: The force is equal to the weight of the apple—a few newtons. The mass of the earth, however, is so great that the resulting acceleration of the earth is insignificant.

The acceleration of the earth is insignificant, but that is no reason to think that it does not exist. There is no reason to think that the law no longer applies, just because its effect is too small for us to measure. By the very nature of force, Newton's laws hold in all cases.

SELF-TEST

1. State Newton's first law. _____

   _____

2. State Newton's second law in equation form. _____

3. Is acceleration a scalar or a vector quantity? _____

4. Three newtons are exerted northward on a 10-kg object while 4 newtons are exerted southward. What is the acceleration of the object? _____

   _____

5. What is the unit of force in the SI system? _____

6. How much does a 3-kg object weigh? _____

7. Explain why a heavier object does not fall faster when dropped, since the accelerating force (its weight) is greater. _____

   _____

8. The acceleration of gravity is 9.8 m/s² downward. In actual practice, does an object continue to accelerate if dropped from a tall building? _____ Explain. _____

9. You have tied a shoe to a rope and are whirling it above your head. Name all forces on the shoe and state the direction of each force. _____

   _____

10. State Newton's third law. _____

    _____

11. My weight is the result of the earth's gravitational force on my body. What is the corresponding reaction force? _____

    _____

ANSWERS

If your answers do not agree with those given below, review the frames indicated in parentheses before you go on to the next chapter.

1. An object at rest tends to stay at rest and an object in motion tends to stay in motion in a straight line at a constant speed. (Or: An object tends to keep the same velocity.) (frame 1)

2. $F = ma$ (frame 2)

3. vector (frames 2, 3)

4. 0.1 m/s² southward (Solution: The resultant of the forces is 1 newton southward. Using $F = ma$, the acceleration a is 0.1 m/s².) (frames 2–4)

5. newton (frames 5, 6)

6. 29.4 N (Solution: If dropped, the acceleration would be 9.8 m/s². Application of Newton's second law yields the force on the object, which is its weight.) (frames 7, 8)

7. Its mass is greater by just the same ratio as its weight. Thus, an application of Newton's second law (**F** = m**a**) results in the same acceleration. (frames 9, 10)

8. no; Because of air resistance, the object reaches a terminal speed. (frames 11, 12)

9. (1) The force of gravity pulls downward on the shoe. (2) The string pulls inward toward the center of the circle. (You might also list air friction, which acts in the direction opposite to the motion of the shoe.) (frames 13, 14)

10. For every force object X exerts on Y, Y exerts an equal and opposite force on X. (frame 15)

11. my body's gravitational force on the earth (frames 16, 17)

3 The Conservation of Momentum and Energy

Prerequisites: Chapters 1 and 2
Conservation laws are an important part of the way the physicist views nature. Two of the most important are the laws of conservation of momentum and of conservation of energy. The former follows directly from Newton's laws of motion, while the latter is much deeper.

OBJECTIVES

After completing this chapter, you will be able to

- define momentum and state its metric units;
- relate momentum to impulse;
- given a change in momentum and the time involved, calculate the average force used;
- explain how momentum is conserved in an isolated system;
- apply the principle of conservation of momentum to analysis of simple collisions;
- calculate velocity after a collision between two objects, given initial velocities and masses of the objects;
- recognize the part played by the earth in conservation of momentum;
- define work and power;
- given force, distance, and time, calculate work and power;
- state correct units for work and power in the metric system;

- define and differentiate between potential energy and kinetic energy;
- calculate potential energy, given weight (or mass) and height;
- explain how potential energy, kinetic energy, and work are related in a system;
- use the principle of conservation of energy to calculate the speed of an object that has fallen from a given height (optional);
- specify how energy conservation applies to machines, giving a quantitative example.

1   MOMENTUM

Momentum is defined as the product of the mass of an object and its velocity. Thus, the SI unit of momentum is kg·m/s. To understand the value of the concept of momentum, let's return to an earlier definition. Acceleration is the change in velocity divided by the time involved, or:

$$\mathbf{a} = \frac{\mathbf{v}_1 - \mathbf{v}_0}{t}$$

Now, recall Newton's second law in equation form:

$$\mathbf{F} = m\mathbf{a}$$

If we substitute for **a** the expression written above, we get:

$$\mathbf{F} = m\left(\frac{\mathbf{v}_1 - \mathbf{v}_0}{t}\right)$$

or

$$\mathbf{F}t = m(\mathbf{v}_1 - \mathbf{v}_0)$$

On the right side of this equation we have the product of mass and change in velocity. This is the change in momentum.

(a) What are the units for the terms of the left side of the equation? _____

_____

(b) What are the units for the right side of the equation? _____

(c) Show that the units on the two sides are equivalent. _____

Answers: (a) newton·second (or N·s); (b) kilogram·meter/second (or kg·m/s); (c) N·s = (kg·m/s²)·s (A newton is a kg·m/s²; thus, N·s = kg·m/s, which is what we have on the right side.)

The product of force and time is defined as **impulse.** The equation $Ft = m(v_1 - v_0)$ shows that impulse is equal to the change in momentum.

Consider what happens when you have a collision in your car. Your body experiences a change in momentum, for while you are moving you have some momentum, but after the collision you have none. Since impulse is equal to the change in momentum, the force that is exerted on your body in order to stop it depends on the time during which that force acts. If you are not wearing seat belts and have no airbag, your head may continue forward until it is stopped *quickly* by the windshield. The equation tells us that if the time (t) is small, the force (**F**) will be great. Seat belts and airbags increase the time during which a person decelerates, therefore reducing the force exerted on the person.

Explain, in terms of impulse and momentum, why it hurts less if you let your hand "give" with a hard-thrown ball when you catch it. _____

Answer: In order to change the ball's momentum to zero, a certain amount of impulse is necessary. If the time during which the ball is stopped is longer, the force will be less.

MOMENTUM CONSERVATION

Consider the two carts in the figure below. There is a compressed spring between them, but a rope is holding them together. Since the carts are in equilibrium, the total force on each of them is zero. If the rope is cut, however, the compressed spring exerts an unbalanced force on each of the objects.

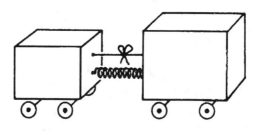

What does Newton's third law (the action–reaction law) tell us about the forces on the carts? _____

Answer: They are equal and opposite.

Since these forces are exerted as long as the spring is in contact with the carts, and since the force is exerted on one cart exactly as long as it is exerted on the other, each cart receives the same impulse. Recall that impulse is equal to the change in momentum. Since each cart receives the same impulse, each cart changes its momentum by the same amount.

There is one difference, however. The forces on the two carts are in opposite directions. Thus, their impulses are in opposite directions, and the direction of their momentum change is opposite. This just means that one gains momentum in one direction but the other gains momentum in the opposite direction.

(a) What is the total momentum of the system before the rope is cut? _____

(b) What is the total momentum of the system while the carts are moving?

_____

Answers: (a) zero; (b) zero (because the carts' momenta (plural of momentum!) are in opposite directions)

5

In the last two frames you have seen an example of the concept of the conservation of momentum. Momentum was "conserved" because the momentum of the carts before the "event" was equal to their momentum after the event. The law of conservation of momentum states:

> The total momentum in any isolated system before any event
> is equal to the total momentum after the event.

There is an important limit to the application of the principle of conservation of momentum: the system must be isolated from other forces. In the example above, if you had pushed on the carts with your hand, they would not form an isolated system.

When a gun fires a bullet, the gun and bullet can be considered an isolated system, at least until the gun recoils back against the shooter's shoulder and the shoulder exerts an outside force on the gun. Before the "event"—the firing of the bullet—the gun and bullet are at rest, so their total momentum is zero. After the bullet is fired, it has momentum. (Although its mass is fairly small, its speed is great, so its momentum is considerable.) If momentum is to be conserved, something must gain momentum in the direction opposite that of the bullet. The gun does just that.

(a) How does the mass of the gun compare to that of the bullet? _____

_____

(b) After the firing, how does the momentum of the gun compare to that of the bullet? _____

(c) Therefore, how does the velocity of the gun compare to that of the bullet after firing? _____

Answers: (a) It is much greater. (b) They are equal. (c) It is much less.

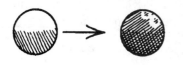 COLLISIONS

The law of conservation of momentum also allows us to analyze collisions between objects. Consider the simple case of one pool ball colliding head-on with another at rest. Before the collision, the total momentum of the system is the product of the mass of the moving ball and its velocity. Those of you who have played pool know what happens after such a collision: the ball that has been moving stops, and the other ball moves forward. The conservation of momentum can tell us what the speed of that second ball must be.

(a) Assuming their masses to be the same, how does the momentum of the second ball after collision compare to that of the first ball before collision?

_____

(b) How does the velocity of the second ball after collision, compare to the initial velocity of the first ball? _____

Answers: (a) They are equal.   (b) They are equal.

7

Consider a wreck in which a car runs into the back of a stationary truck. Before the collision, the car has momentum forward. We will suppose that the car and truck stick together after the crash.

Before

After

The law of conservation of momentum tells us what the momentum of the wreckage will be.

(a) What will it be equal to? _____

(b) What values would we need to know to calculate the velocity of the wreckage after impact? _____

_____

Answers: (a) the momentum of the car before collision;   (b) masses of car and truck, and pre-wreck velocity of the car

8   Since momentum is the product of a vector and a scalar, momentum is also a vector. To see what effect this has, let's go back to the pool table. In part 1 of the figure, ball A is about to collide with ball B again, but it is coming in a direction such that they will not hit straight on. You would expect the balls to continue somewhat as shown in part 2. Before the collision, the total momentum of the system was in a downward direction on the page. After collision, the momentum of ball A is not only downward, but also toward the right.

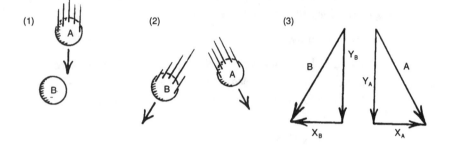

If ball A moves toward the right, what does the law of conservation of momentum tell us about the motion of ball B after the collision? _____

_____

Answer: It has an equal momentum toward the left.

9   In part 3 of the previous figure, arrow X_A, which represents the sideways momentum of A after collision, must be the same length as arrow X_B, representing ball B's sideways momentum. Vectors Y_A and Y_B represent the downward momenta of balls A and B.
    The total of these two vectors must equal what momentum? _____

_____

Answer: the momentum of A before collision

10 # MOMENTUM PROBLEMS

A gun has a mass of 10 kilograms and its bullet has a mass of 20 grams. Suppose that the bullet fires out of the gun barrel with a speed of 300 m/s (about 600 miles per hour). What is the recoil speed of the gun? _____

Answer: 0.6 m/s
Solution: ("G" stands for "gun" and "b" for "bullet")
Since the initial momentum = 0, the total final momentum is zero:

$m_G\mathbf{v}_G + m_b\mathbf{v}_b = 0$
(10 kg)·$\mathbf{v}_G$ + (0.02 kg*)·300 m/s = 0
$\mathbf{v}_G$ = −0.6 m/s

(The negative answer tells us that the velocity is opposite in direction to the 300 m/s.)

11   A ball of mass 200 grams (a softball, perhaps) is moving at 3 m/s before it collides head-on with a baseball of mass 150 grams. After collision the baseball is
moving forward at a speed of 2.5 m/s. How fast is the softball moving after collision? (Hint: You may use grams as your unit of mass here as long as you use it
on both sides of the equation.) _____

Answer: 1.125 m/s

If you failed to get the correct answer, try again with the following hint before
checking the solution below. Hint: The total momentum after collision is equal to
the total momenta of the two balls before collision.

Solution:

$m\mathbf{v}_{before}$ = $m\mathbf{v}_{after}$
(200 g)·3 m/s = (200 g)·$\mathbf{v}$ + (150 g)·(2.5 m/s)
600 g·m/s = (200 g)·$\mathbf{v}$ + 375 g·m/s
225 g·m/s = 200 g·$\mathbf{v}$
$\mathbf{v}$ = 1.125 m/s

12 FRICTION AND THE EARTH

We have thus far neglected to mention the effect of friction. We spoke of the
momentum of cars and trucks immediately before and after collisions, for example, but said nothing of the fact that after a wreck the wreckage would slide to a
stop. What happens to the momentum of the wreckage? The answer is that the
momentum is transferred to the earth. The earth spins a little faster (or slower,
depending upon whether the motion was eastward or not). It may at first glance
seem hard to believe that an object as small as a truck can affect the speed of the
earth, but one must remember that we are not saying that the effect would be
great enough to measure. Since the mass of the earth is so tremendous, its change
in speed would be very, very small. Undetectable—but not nonexistent.

---

*Note that mass must be expressed in the same units on each side of the equation. You can't use
"20 g" unless you converted 10 kg to 10,000 grams.

Consider a recoiling gun when it hits the shooter's shoulder. The system is not isolated, because it includes the entire earth. Is the change in momentum of the shooter detectable? _____

Answer: Perhaps (he may fall over backwards). If the shooter is ready, however, he braces against the expected impulse and the momentum is transferred to the earth.

13 WORK

In physics, work is defined as the product of the net force and the displacement through which that force is exerted, or W = **Fd**. This corresponds to our everyday meaning of the word in that when you lift a brick 6 feet from the floor you do twice as much work as when you lift it 3 feet. Or suppose you lift 50 bricks to a height of 3 feet. You do 50 times as much work as lifting one brick that high.

But here comes a conclusion that may seem odd. After you lift a dozen bricks up to your waist, you are told by your boss to hold them there for 5 minutes. Do you do work in holding them there? To answer this we must go back to our definition. The force you exert is simply the weight of the 50 bricks.

(a) In holding them waist-high, does that force move the bricks through a distance?

_____

(b) Using the formula, calculate the total work done in holding the bricks. _____

Answers: (a) no; (b) zero

(In holding the bricks, there is motion within the muscles, so work is done inside your body, but no external work is done on the bricks. It is this external work with which we are concerned.)

14

If force is measured in pounds and distance in feet, the unit of work is the ft·lb, or foot·pound. In SI units, recall (from frame 5 of Chapter 2) that force is measured in newtons and distance in meters. The unit of work is the newton·meter, called the **joule** (pronounced "jewel").

The defining equation for work, W = **Fd**, indicates that force and displacement are vectors, but that work is not. The vector nature of force and displacement requires that the force and displacement must be in the same direction if they are to be multiplied to obtain work.

(a) How much work is done in lifting a 3-newton cabbage 2 meters? _____

(b) How much work is done in carrying a 12-newton hammer while holding it at a constant 1.5 meters above the floor? _____

Answers: (a) 6 joules; (b) zero

(The reason that no work is done in part (b) is that as you carry the hammer you exert a force upward on it, but there is no displacement in this direction. You do not exert a force horizontally while you are moving something horizontally at a constant speed.)

15 POWER

Power is defined as work done divided by the time used to do the work, or $P = W/t$. Here again an everyday term is defined in a way that makes it measurable. In a given amount of time a powerful engine can do more work than an engine with less power.

If two engines do the same amount of work, what advantage is there to using the more powerful engine? _____

Answer: The more powerful one can do the work in less time.

16

The SI unit of power is the joule/second. Just as we renamed the newton·meter as the joule, we define 1 joule/second as 1 **watt**. An example: A particular machine is capable of a power of 1000 watts. How far can it lift a 50-newton weight (about 11 pounds) in 3 seconds? To answer this, combine the work and power formulas to get

$$P = \frac{Fd}{t}$$

(a) Use your knowledge of algebra to find the formula for **d**.   d = _____
(b) Solve the example problem. Your answer? _____

Answers: (a) $\dfrac{Pt}{F}$;   (b) 60 meters

17 POTENTIAL ENERGY

Energy is defined as the ability to do work. There are many types of energy, including electrical energy, heat, and nuclear energy. In this chapter we will be concerned primarily with two types of energy: gravitational potential energy and kinetic energy.

An object can have energy—the ability to do work—because of position. A weight that is high above your head can be made to exert a force when it falls. Since gravity is the ultimate source of this energy, it is correctly called **gravitational potential energy**, but we usually abbreviate this to "potential energy."

Consider the pendulum in the figure. Release it. When the pendulum hits the shoe, it will push the shoe forward, exerting a force on the shoe over a distance. If we were to multiply the force exerted on the shoe by the distance the shoe moves, we could calculate the amount of work that is done.

The pendulum had the ability to do work on the shoe because when the pendulum was at the raised position it had what type of energy? _____

Answer: gravitational potential energy

18

An object at a given height has an amount of potential energy (PE) that is equal to the work done in raising it to that height. Or, since the force to lift an object is equal to the object's weight,

$$PE = weight \cdot height$$

Consider an 8-newton object that is 2 meters above the floor.

(a) What is its potential energy with reference to the floor? _____

(b) How much work can it do when it falls? _____

(c) Suppose the object falls onto a marshmallow on the floor and mashes the marshmallow down 1 centimeter. What force is exerted by the object on the marshmallow? _____

Answers: (a) 16 joules;   (b) 16 joules   (c) 1600 N

Solution to part (c):
$$W = \mathbf{Fd}$$
16 joules $= \mathbf{F} \cdot 0.01$ m
$$\mathbf{F} = 1600 \text{ newtons}$$

19

Now suppose that a 5-kg object rests on a table. You lift it up to the ceiling, which is 2 meters above the table. What is its potential energy with reference to the table? (You can calculate this immediately, or follow the steps below.)

_____

Remember that 5 kg is the mass of the object, and that we must find its weight, since its weight is equal to the force we have to exert to lift it. Newton's second law states that

$$F = m\mathbf{a}$$

or in this case:

Weight = $m\mathbf{g}$ (where $\mathbf{g}$ is the acceleration of gravity)
Weight = 5 kg (9.8 m/s²)
Weight = _____ (include units in your answer)
PE = weight·height
PE = _____

Answers: Weight = 49 newtons, since 1 kg (m/s²) is a newton; potential energy = 98 joules (or newton·meters).

20   KINETIC ENERGY

The pendulum shown here is just about to hit the shoe. When it was at position A, it had potential energy. Just before it hits the shoe, it has fallen to the level of the shoe and has thus lost its potential energy. So how can it move the shoe if it has no potential energy anymore? The answer is that the pendulum still has energy, but now the energy is due to its motion rather than to its position. **Kinetic energy** is the energy an object has by virtue of its motion.

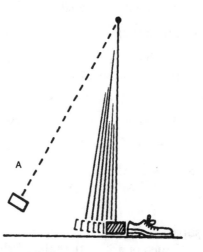

(a) What form of energy is due to position? _____

(b) What form of energy is due to motion? _____

(c) Which of these energy forms can a pendulum have? _____

Answers: (a) potential; (b) kinetic; (c) both potential and kinetic

21 The amount of kinetic energy an object has depends upon the object's mass as well as its speed. Mass, rather than weight, is the important factor because even if the object were far enough from a planet to be considered weightless, it would still have kinetic energy if it were moving. The formula we use to calculate kinetic energy (KE) is:

$$KE = \frac{1}{2}mv^2$$

(a) Would doubling the mass or doubling the velocity of an object have a greater effect on its kinetic energy? _____

(b) Suppose the velocity of an object is tripled. How does its kinetic energy change? _____

Answers: (a) velocity (since it is squared in the formula); (b) It would be nine times as great ($3^2 = 9$).

22 ENERGY CONSERVATION

A pendulum will always rise to the height from which it fell, unless—as when it hit the shoe in the previous example—it does work on something. If a pendulum does not hit anything as it swings, it rises to very nearly to the same height as that from which it was released. The reason that it does not quite reach the height from which it was released is that the pendulum does work on something as it moves—it does work on the surrounding air. The pendulum exerts a force on air molecules and pushes them aside. But if a pendulum is not moving too fast, it does not exert a great amount of force on the air molecules, and it rises very nearly to its original height. Let's consider the potential energy of the pendulum.

Imagine a pendulum with no air friction. In this case the pendulum will rise to its original height. How does the potential energy at one end of the swing compare to the potential energy at the other end? _____

Answer: They are equal.

Therefore, we say that potential energy is **conserved** in this case.

23 Consider the pendulum that hits a shoe at the bottom of its motion. The kinetic energy of the pendulum just before it hits the shoe is equal to the potential energy it lost in falling from the left end of its swing. In fact, if we ignore air friction and if the shoe were not there, the potential energy attained on the right end of the swing would be equal to the kinetic energy at the bottom. Now consider the effect of friction, and compare the potential energy at the right end of the swing to the potential energy originally at the left.

(a) If friction is present, how will the potential energy at the right compare to that on the left? _____

(b) What will cause this? _____

Answers: (a) It will be less.   (b) work done on the air molecules

24

The conservation of energy applies during the entire fall of an object. At any time, the decrease in potential energy is equal to the increase in kinetic energy. Stated another way, the total of the kinetic and potential energy is constant. We will see later in the book that the law of conservation of energy includes all types of energy. The law of conservation of energy can be stated as follows:

> The total energy at the end of any event is equal to
> the total energy before the event.

Or

Energy can be neither created nor destroyed.

(If you are concerned with the energy that is "lost" when there is friction, we will see that when this occurs, heat, a form of energy, is produced. So energy is indeed conserved.*)

(a) At the point where a pendulum has lost 9 joules of potential energy, how much kinetic energy has it gained? _____

(b) How much potential energy is needed to develop 15 joules of kinetic energy?

_____

(c) Suppose the pendulum has 20 joules of PE at the top of its swing. When the pendulum is partway down, its PE is 12 joules. How much KE does it have at this point? _____

Answers: (a) 9 joules;   (b) 15 joules ;   (c) 8 joules

---

*Note that we use the word *conservation* in a different sense than it is often used in everyday language. The law of conservation of energy does not refer to the fact that you and I should not waste energy in activities, but it means that energy cannot be destroyed. However, energy can be changed from one form to another, and those exchanges are what we are studying here.

25   ENERGY IN COLLISIONS

Earlier in this chapter we used the law of conservation of momentum to make predictions of velocities after collisions. Where does the concept of energy fit in here? In the simplest collision, in which a pool ball hits a second one at rest and the second one assumes the velocity of the first, there is no kinetic energy lost. Such a collision is said to be **perfectly elastic**. Most collisions are not perfectly elastic; there is less kinetic energy after a collision than before. The law of conservation of energy is not violated in such **inelastic** collisions, though, because the temperature of the objects increases, indicating an increase in heat energy.

When two cars collide, is the collision normally elastic or inelastic?

———————————

Answer: inelastic

26   MORE ENERGY CALCULATIONS (OPTIONAL)

A man drops a 10-kg rock from the top of a 5-meter ladder. Assuming that air friction is negligible, how much kinetic energy does it have when it hits the ground? Calculate the answer yourself if you can. Otherwise use the steps below. (Hint: First calculate the potential energy at the top.)

Step-by-step solution: Potential energy at top = weight·height

$$PE = mg \cdot \text{height}$$

(Remember, weight = mg, from Chapter 2, frames 2, 7–10.)

(a) PE = ____ kg·____ m/s^2·____ m

(b) PE = _____ (kg·m/s^2)·m

(c) PE = _____·m

(d) PE = _____

Answers: (a) 10, 9.8, 5;   (b) 490;   (c) 490 newton;   (d) 490 joules

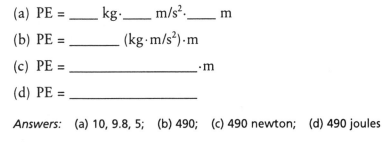

27

What is the speed of the rock (in the previous example) just before it hits the ground? _____

Answer: 10 m/s (Since we are assuming that no energy was lost to frictional forces, the potential energy at the top is equal to the kinetic energy at the bottom.)

$$KE = \frac{1}{2}mv^2$$
$$490 = \frac{1}{2} \cdot 10 \cdot v^2$$
$$v^2 = 98$$
$$v \cong 10 \text{ m/s}$$

($\cong$ means "approximately equal to")

SELF-TEST

1. How are impulse and momentum related? _____

    _____

2. How does the momentum of a bullet as it leaves the barrel of a gun compare to the recoil momentum of the gun? _____

3. Suppose a moving object collides with a stationary one and the two stick together. How will the speed of the stuck-together objects compare to the speed of the original moving object? _____

4. Is momentum a scalar or a vector quantity? _____

5. Two blocks of ice are sitting on a table with a piece of dynamite between them. One block of ice has a mass of 12 kg and the other's mass is 18 kg. The dynamite explodes. If the block of lesser mass moves off with a speed of 10 m/s, what is the speed of the more massive block? _____

6. A 2000-pound car moving at 50 miles per hour collides with a stationary car weighing 3000 pounds. If the two cars stick together, what is their speed immediately after collision? (Hint: You may use "pounds" for mass and "miles per hour" for velocity, as long as you do so on both sides of the equation. This is permissible because weight (pounds) is proportional to mass, and miles/hour is a legitimate unit of velocity.) _____

7. Is the law of conservation of momentum violated if a gun is attached so firmly to a support that it cannot recoil? Explain. _____

    _____

8. How is work defined? _____

    _____

9. How much work is done in holding a piano at shoulder height for 15 minutes if the piano weighs 1200 newtons and your shoulder is 1.5 meters above the ground? _____

10. How is power defined? _____

_____

11. As an object falls toward the ground, what happens to its potential energy?

_____

12. Suppose two objects have the same mass, but one is moving twice as fast as the other. The kinetic energy of the fast one is how many times as great as the kinetic energy of the slow one? _____

13. What happens to the kinetic energy of an object as the object falls?

_____

14. How does a perfectly elastic collision differ from an inelastic collision?

_____

15. *Optional.* Lifting a 5-kg object 3 meters requires how much energy?

_____

16. *Optional.* If the object in question 10 is lifted in 2 seconds, what is the power of the lifter? _____

17. *Optional.* An object is going 10 m/s when it hits the ground. From how high did it fall? (Hint: Since the object's mass does not matter, you may choose any mass you want.) _____

ANSWERS

1. The impulse applied to an object is equal to the change in momentum of the object. (frames 1, 2)

2. The two momenta are equal (but opposite in direction). (frames 3–5)

3. It will be less. (The momentum of the stuck-together objects must be equal to the momentum of the original moving object. Since the two objects have a mass greater than the mass of the single object, the speed of the two objects must be less.) (frames 6, 7)

4. vector (frames 8, 9)

5. 6.67 m/s. Solution: The total momentum before the explosion is zero, so the momentum of one piece of ice after the explosion plus the momentum of the other piece must equal zero.

$$12 \text{ kg} \cdot 10 \text{ m/s} + 18 \text{ kg} \cdot \mathbf{v} = 0$$
$$\mathbf{v} = -6.67 \text{ m/s}$$

(The minus sign tells us that the speed of this block is in the opposite direction from the other block.) (frame 10)

6. 20 miles/hour. Solution:

$$(2000 \text{ lb} \cdot 50 \text{ mi/hr}) + (3000 \text{ lb} \cdot 0) = 5000 \text{ lb} \cdot \mathbf{v}$$
$$100{,}000 \text{ lb} \cdot \text{mi/hr} = 5000 \text{ lb} \cdot \mathbf{v}$$
$$\mathbf{v} = 20 \text{ mi/hr}$$

(frame 11)

7. No. The support—and the entire earth—recoils. (frame 12)

8. Work is the product of the force exerted and the distance through which it is exerted. (frame 13)

9. No work is done. (frame 14)

10. Power is work done divided by the time used in doing the work. (frame 15)

11. It decreases (changing into kinetic energy and perhaps heat). (frames 17–19)

12. four times (from the formula $KE = \frac{1}{2}mv^2$ (frame 21)

13. It increases (unless there is so much friction that the object does not gain speed). (frames 22, 23)

14. In a perfectly elastic collision, kinetic energy is conserved, but in an inelastic collision it is not. (frame 25)

15. 147 joules (Remember, you must use the *weight* of the object, and the weight is equal to the mass times the acceleration of gravity.) (frame 18)

16. 73.5 watts (147 J/2 s = 73.5 J/s) (frames 15, 18)

17. 5.1 meters ($m\mathbf{g}h = \frac{1}{2}mv^2$, thus $\mathbf{g}h = \frac{1}{2}\mathbf{v}^2$, and $h = \mathbf{v}^2/2\mathbf{g}$) (frame 27)

4 Gravity

Prerequisite: Chapters 1 and 2

The law of gravitation plays a very prominent role in our view of the universe. The force of gravity causes a pin to fall to the floor, holds our planet in its orbit around the sun, holds the sun and some 200 billion other stars in our Milky Way galaxy, and causes the gigantic galaxies to cluster together in groups. Yet the law of gravity was unknown to humans until rather recently in history. And here again the name of Isaac Newton appears. He is the genius who first hypothesized that the heavens and the earth are united by this single, all-encompassing principle.

OBJECTIVES

After completing this chapter, you will be able to

- state the law of gravity using words or a formula;
- explain how gravity keeps the moon in orbit;
- state the shape of a planet's orbit and the sun's position with respect to that orbit;
- give examples to illustrate the effect of distance on the gravitational force between two objects;
- explain how the law of equal areas defines the speed of a planet around the sun;
- explain how Newton's laws account for the law of equal areas;
- specify the effect on orbiting satellites of the force of gravity;
- justify the concept of "apparent weightlessness";
- use numerical examples to illustrate how the acceleration of gravity varies with height above the earth;
- describe the relationship between Kepler's laws and Newton's laws;
- give the value of the universal gravitational constant (optional);

• calculate the gravitational attraction between two objects, given masses and distances (optional);

• use the law of gravity to calculate the mass of the earth (optional).

1 THE LAW OF GRAVITY

Newton's law of gravity can be stated as follows:

Every object in the universe attracts every other object in the universe with a force proportional to the objects' masses and inversely proportional to the square of the distance between their centers.

The statement is much shorter in symbolic form:

$$\mathbf{F} \sim \frac{m_1 \cdot m_2}{d^2}$$

The symbol ~ means "is proportional to"; m_1 symbolizes the mass of one of the objects, and m_2 is the mass of the other; d is the distance between the centers of the objects.

The law says that there is a force of gravity between *every* two objects, between you and the person next to you, and between you and a bus out on the street. Why do you not feel the force of gravity between you and this book?

Answer: The masses involved are so small that the force is very small.

2

Consider how the force of gravity between two objects changes as the distance between them changes. An **inverse square relationship** is involved—the force changes inversely as the square of the distance. Thus, if the distance between the two objects is tripled, the force becomes $\frac{1}{9}$ as great ($1/3^2 = 1/9$).

(a) If the distance is made four times as great, the force of gravity is what fraction of its original value? _____

(b) Suppose two objects are initially 5 meters apart and then are moved to 1 meter apart. How does the force of gravity between them change? State your answer mathematically. _____

Answers: a) $\frac{1}{16}$;   (b) It becomes 25 times greater.

3

The distance involved in the law of gravity is the distance between the centers of the two masses. My distance to the center of the earth is about 4000 miles when I am on the surface.

(a) If you go to a height of 4000 miles above the surface of the earth, the force of gravity (your weight) then becomes what fraction of your present weight?

_____

(b) Suppose you weigh 117 pounds here on earth. How far from the center of the earth must you go to weigh 13 pounds? _____

Answers: (a) $\frac{1}{4}$ (you are then 8000 miles—twice as far—from center);   (b) 12,000 miles

Solution to part (b): The ratio of the forces is $\frac{13}{117}$, or $\frac{1}{9}$. To decrease the force to $\frac{1}{9}$, you must be 3 times as far away.

4

In Chapter 1 you learned that the acceleration of gravity is 32 ft/s^2 or 9.8 m/s^2. This is true only near the surface of earth, however, because at different distances from earth-center the force of gravity is different. Over the surface of the earth the difference in distance is slight, and the acceleration due to the force of gravity ranges from about 9.788 m/s^2 to about 9.808 m/s^2.

We have seen that when an object is dropped from rest, the distance it falls in a given time depends directly on the object's acceleration ($d = \frac{1}{2}at^2$—Chapter 1). We have also seen that the acceleration varies directly as the net force applied to the object (F = ma—Chapter 2). Thus, the distance fallen in 1 second is directly proportional to the force of gravity. This means that an object that is 4000 miles above the surface of earth (where the force of gravity is $\frac{1}{4}$ as great) falls 4 feet instead of 16 feet during the first second of fall.

(a) As one rises above the surface of earth, does the acceleration of gravity become greater or less? _____

(b) Does an object fall more or less far during the first second of fall if it is high above the earth compared to being at sea level? _____

(c) Suppose an object is 60 times as far from earth-center as the earth's surface is from earth-center. What fraction of 4.9 meters (or 16 feet) will the object fall in one second? _____

Answers: (a) less;   (b) less far;   (c) 1/60^2 (or 1/3600)

5 THE MOON AND GRAVITY

The answer to the last question applies to any object, including the moon (which, in fact, is 240,000 miles from earth-center, or 60 times as far as we are from center). Consider the moon moving in a direction shown by the solid arrow in the figure. The moon would continue in a straight line if no force were applied to it.

In fact, a force of gravity is exerted on it by the earth, so instead of going in a straight line, the moon's path is bent by the force. In 1 second, the moon "falls" from a straight-line path by just the amount predicted by Newton's law of gravity, $1/60^2$ of 4.9 meters. If you do the calculations, you will find that this distance is 0.00136 meters, or 0.136 centimeters. Isaac Newton did this calculation and realized that it confirmed his law of gravity. It was the first application of a law of nature to both heavenly objects and earthly ones.

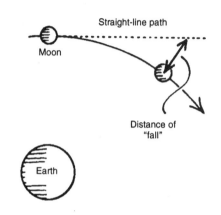

(a) Which of Newton's laws says that the moon should go in a straight line?

_____

(b) Which law explains why it doesn't? _____

Answers: a) first law;   (b) law of gravity

6 KEPLER'S LAWS

About 50 years before Newton formulated the law of gravity, Johannes Kepler (1571–1630) had found some rules that could be used to calculate the orbits of the planets. The reason that these rules worked was not known until Newton applied his gravitational theory to planets, however. We will consider Kepler's laws and see how they are explained by the laws of Newton.

According to Kepler's first law, planets orbit the sun in the shape of an ellipse. An ellipse is the shape that a circle appears to be when one looks at it from an angle. The easiest way to draw an ellipse is to put two tacks in a piece of paper and put a loose loop of string around them, as shown in the first part of the figure that follows. Now take a pencil and stretch the string out. Then sweep around the tacks, always keeping the string taut. The figure drawn is an ellipse, and the two points where the tacks are located are called the **foci** (the plural of **focus**) of the ellipse. Kepler's first law further states that the sun is located at one focus of the elliptical path of the planet.

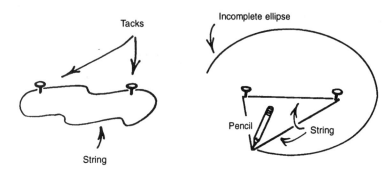

(a) If the tacks are placed farther apart, how will the shape of the ellipse be changed? _____

(b) What would the positions of the tacks have to be to draw a circle?

_____

(c) State Kepler's first law. _____

Answers: (a) It will be flatter.   (b) both at the same point (Thus, a circle is just a special form of an ellipse.);   (c) Planets orbit the sun in elliptical paths with the sun at one focus of the ellipse.

7   Kepler's second law:

As a planet orbits the sun, a line from the planet to the sun sweeps across equal areas in equal times.

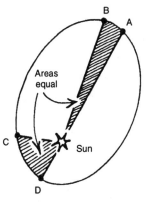

To see what this means, refer to the figure. Suppose that the earth moves from point A to point B in a 2-week period. Later in the year, it is at point C, closer to the sun. The law tells us that during the next 2 weeks it will go just far enough that the area swept out by an imaginary line to the sun will be equal to the area it swept out when it moved from A to B.

Does the planet move faster between A and B or between C and D? _____

Answer: between C and D

8   NEWTON'S EXPLANATION FOR KEPLER'S SECOND LAW

Now we'll see how Newton's laws explain these apparently arbitrary rules found by Kepler. In the next figure, the force of the sun's gravity on the planet is represented by an arrow. Note that although the arrow is not pointed in the direction the planet is moving, it is not pointed at a right angle to that direction either. The force of gravitation toward the sun will have two effects on the planet's motion: It will cause the path to curve, and it will cause the planet to increase its speed (since the force is pulling forward as well as to the side).

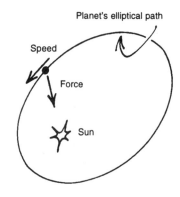

Next, consider the motion of the planet when it is halfway around its orbit from where it is shown in the figure. In this case the force will still cause the planet's path to curve, but now the force will also slow down the planet.

If you do the necessary math, you will find that the amount of speedup or slowdown of the planet would be just the right amount to give the results that Kepler stated as his law of equal areas. This is a very important point: Kepler's laws were simply concise statements of how planets moved, but they did not relate planet motion to any other law of nature. Newton's laws apply to objects here on the surface of the earth as well as in space.

(a) What is the shape of the earth's orbit around the sun? _____

(b) Does the earth move at a constant speed around the sun? _____

Answers: (a) an ellipse;   (b) no

9 KEPLER'S THIRD LAW (OPTIONAL)

Kepler's third law states the relationship between a planet's average distance from the sun and the time required for the planet to complete one orbit. The time for one cycle of the sun is called a planet's **period** of revolution. The relationship can be stated as follows, where R (for radius) is the distance from the planet to the sun and T (for time) is the period of revolution:

$$T^2 \sim R^3$$

Newton's laws explain this law of Kepler in that planets at greater distances experience less force from the sun and, if they are to orbit at that distance, they must be moving more slowly than planets closer to the sun. A mathematical application of Newton's laws results exactly in Kepler's third law.

(a) Define the period of a motion. _____

(b) Do planets at greater distances from the sun have longer or shorter periods of revolution? _____

Answers: (a) the time to complete one full cycle;   (b) longer

10 EARTH SATELLITES

This figure is similar to one found in Newton's book, *The Principia*. Newton argued that if a cannon could be placed on a hill high enough to be outside the earth's atmosphere, and if the cannon could be made powerful enough, a cannonball could be placed in orbit around the earth.

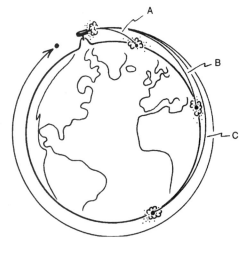

First, suppose that the shot from the cannon results in a path like that of path A in the figure. If the cannon were more powerful, the ball might take a path like B, or even C. And finally, if a powerful enough cannon could be obtained, it could shoot the cannonball so that the cannonball came back to where it started. It would then be in orbit.

(a) What keeps such a cannonball from flying into space? _____

(b) Why does the cannonball never reach the ground? _____

Answers: (a) the force of gravity;   (b) The curvature of the earth is just enough so that as the cannonball falls, it gets no closer to the surface.

11 WEIGHTLESSNESS

Heavy and light objects fall at the same rate. Thus, if a baseball were fired side-by-side with the cannonball they would "fall" side-by-side around the earth. In fact, if the cannonball were hollow and had a baseball and a penny inside, the baseball and penny would orbit with the cannonball without pushing against the inside of the ball. They would seem **weightless**. But are they truly weightless? To answer this, recall that we define weight as the force of gravity, and the force of gravity is the force that keeps the cannonball, baseball, and penny from going off into space. Thus, according to our definition of weight, objects in orbit are not truly weightless.

An astronaut in orbit around the earth is "falling," in the same sense that the cannonball is "falling." She seems to float weightlessly around in the orbiting ship. She is not, however, beyond the pull of gravity and is therefore not truly weightless. We should call her condition one of **apparent weightlessness**. Let's check to see how the pull of gravity in orbit compares to the pull of gravity on earth.

(a) If an astronaut were to orbit at the height of 4000 miles above the surface, what would be the astronaut's weight compared to her weight on earth?

_____

(b) Our manned spacecraft orbit only a few hundred miles above the surface, however. In such nearby orbits, the pull of gravity is still 80 to 90 percent of what it is on the surface. Is the astronaut's weight more nearly equal to what it is on earth or to the 4000-mile figure? _____

Answers: (a) $\frac{1}{4}$ of her "normal" weight ;   (b) much closer to her earth weight

12 CALCULATIONS (OPTIONAL)

Earlier we stated the law of gravity in symbols that included a symbol for proportionality. In order to use this relationship for solving problems, we must remove the proportionality by placing a constant in the expression:

$$F = G \cdot \frac{m_1 \cdot m_2}{d^2}$$

The constant G is known as the **universal gravitational constant**. The value of G is:

$$G = 6.67 \cdot 10^{-11} \frac{N \cdot m^2}{kg^2} *$$

(Force is measured in newtons, distance in meters, and mass in kilograms.) Calculate the gravitational attraction between two people standing 1 meter apart. Take the mass of each person as 70 kg. (This is about the mass of a 155-lb person.)

$F = $ _____

Answer: $3.3 \cdot 10^{-7}$ newtons.

Solution: $F = 6.67 \cdot 10^{-11} \frac{N \cdot m^2}{kg^2} \cdot \frac{70 \, kg \cdot 70 \, kg}{(1 \, m)^2}$

Recalling that a newton is about $\frac{1}{4}$ of a pound, we see that this is less than 10^{-7} lb, or less than one ten-millionth of a pound! No wonder we do not experience it!

13   Historically, the gravitational force was carefully measured between two "ordinary" objects with masses such as in the last frame. This allowed a calculation of

_____

*See Appendix I for an explanation of powers-of-ten notation.

the constant G. Once G was known, the mass of the earth—until then unknown—was calculated. You can do it, too.

On the surface of the earth we are $6.4 \cdot 10^6$ meters from the center. In Chapter 2 we learned that a 1-kilogram object weighs 9.8 newtons. To use these facts to calculate the mass of the earth, let m_1 be the earth's mass.

(a) Solve the equation for $m_1 \cdot m_1 =$ _____

(b) Calculate m_1, the mass of the earth. $m_1 =$ _____

Answers: (a) $m_1 = \dfrac{F \cdot d^2}{m_2 \cdot G}$ (**F** is the weight of the object with mass m_2.); (b) $m_1 = 6 \cdot 10^{24}$ kg

Solution: $m_1 = \dfrac{9.8\,\text{N}(6.4 \cdot 10^6\,\text{m})^2}{1\,\text{kg} \cdot 6.67 \cdot 10^{-11}\,\text{N} \cdot \text{m}^2/\text{kg}^2}$

SELF-TEST

1. State the law of gravity, including the dependence of the force on the masses and distance involved. _____

_____

2. How far must a person rise above the surface of the earth (which is 4000 miles from the center of earth) in order to decrease his or her weight to $\frac{1}{4}$ the amount it is on the surface? _____ To $\frac{1}{25}$ of his or her surface weight? _____

3. If the earth were twice as massive, but still the same size, how would our weight differ? _____

4. Use Newton's laws of motion and gravity to explain the fact that the moon's path continually curves toward the earth. _____

_____

5. What is Kepler's first law (the law of ellipses)? _____

_____

6. Both Kepler's and Newton's laws accurately describe the paths of the planets. What advantage do Newton's laws have over Kepler's? _____

_____

7. Use Newton's laws to explain why a planet speeds up during part of its orbit.

_____

8. What holds a satellite up? Discuss. _____

_____

9. Are astronauts in orbit truly weightless? Explain. _____

_____

10. *Optional.* How much gravitational force is exerted between two 1-kilogram masses 1 meter apart? _____ Two meters apart?

_____

11. *Optional.* The mass of the moon is about $7.5 \cdot 10^{22}$ kg, and its diameter is about $3.5 \cdot 10^6$ meters. Calculate the weight of a 70-kg person on the moon. _____

_____

ANSWERS

1. Every object in the universe attracts every other object with a force proportional to the objects' masses and inversely proportional to the square of the distance between their centers. Or:

$$\mathbf{F} \sim \frac{m_1 \cdot m_2}{d^2} \quad \text{(frame 1)}$$

2. 4000 miles; 16,000 miles. Solution: One must get twice as far from earth's center in order to decrease one's weight to $\frac{1}{4}$ the surface value. This is 4000 miles above the surface. To decrease weight to $\frac{1}{25}$, one must move 5 times as far away as the surface. This is 20,000 miles from center, or 16,000 miles from the surface. (frames 2–4)

3. twice as much (frames 1–4)

4. The moon would move in a straight line in the direction it was moving when gravity was "turned off." If there were no unbalanced forces acting on the moon, it would continue in a straight line. The force of gravity acts to change its direction of motion toward the earth. (frames 5, 6)

5. A planet orbits the sun in an elliptical path with the sun at one focus of the ellipse. (frame 6)

6. Newton's laws also apply here on earth, while Kepler's laws seem to have no relation to motions on earth. (frame 8)

7. As the planet is moving along its ellipse in a direction toward the sun, the effect of the force of gravity is to speed up the planet as well as curve its path. (frame 8)

8. Nothing. The satellite is falling as it rounds the earth, in the sense that it does not continue in a straight line as something would that is not affected by gravity. (frame 10)

9. Not if weight is defined as the force of gravity. The astronauts are in free fall along with their space-ship. They fall together the same as two coins fall side-by-side when dropped. (frames 10, 11)

10. $6.67 \cdot 10^{-11}$ N; $1.67 \cdot 10^{-11}$ N.
Solution:

$$\mathbf{F} = G \cdot \frac{m_1 \cdot m_2}{d^2}$$
$$= 6.67 \cdot 10^{-11} \frac{\text{N} \cdot \text{m}^2}{\text{kg}^2} \cdot \frac{1\,\text{kg} \cdot 1\,\text{kg}}{(1\,\text{m})^2}$$
$$= 6.67 \cdot 10^{-11}\,\text{N}$$

At 2 meters distance, the force will be $\frac{1}{4}$ of this, or $1.67 \cdot 10^{-11}$ N. (frame 12)

11. 114 N.
Solution: Substitute into the gravitational equation, being careful to use the radius of the moon rather than the diameter. (frame 12)

5 Atoms and Molecules

No Prerequisites

Ancient humans must have looked at the countless substances in their everyday experience and wondered if those things were perhaps made up of a smaller number of more fundamental, elementary substances. Today we know that this is indeed true—the entire physical world is made up of a relatively few substances, and these exist as atoms and molecules. The basic nature of atoms and molecules affects almost every aspect of physics. In this chapter we take a trip to the unimaginably small world of atoms.

OBJECTIVES

After completing this chapter, you will be able to

- differentiate between elements and compounds;
- specify the total number of elements now known;
- differentiate between atoms and molecules;
- relate atoms and molecules to elements and compounds;
- give an example of the application of the Law of Definite Proportions;
- identify the relative number and type of atoms in a compound, given its formula and the periodic table;
- label the locations of the atomic particles on a drawing of the Bohr model of the atom;
- identify the component of an atom that determines its chemical properties;
- define atomic mass and atomic number;
- interpret portions of the periodic table;
- specify, using an example, the importance of the number of electrons in the outer orbit of an atom;

- use Avogadro's number in describing the number of atoms and/or molecules in a given number of moles of a substance;

- given the number of grams of a compound and the periodic table, calculate the number of molecules of that compound.

1   ELEMENTS AND COMPOUNDS

All of the materials and substances that we experience are made up of 100 or so **elements**, which are the substances into which all matter can (in principle) be divided. Examples of some of the elements are hydrogen, oxygen, carbon, iron, silicon, sulfur, and uranium. Table 5.1 (on facing page) is a complete list of the elements known today and their abbreviations.

Each element has certain characteristic properties. For example, hydrogen is a gas at ordinary temperatures, but it liquifies at a known (lower) temperature. It has a certain density for a given temperature and pressure, and it forms certain compounds when it combines with other elements. Any material which, when tested, has all the properties of hydrogen must be hydrogen. If any properties are different, it must be something other than hydrogen.

Most substances we deal with daily are not elements but **compounds**—chemical combinations of two or more elements. Water, for example, is composed of hydrogen and oxygen. Carbon dioxide is composed of carbon and oxygen. Refer to Table 5.1 as needed to identify each of these as an element or compound.

(a) water_____

(b) alcohol _____

(c) carbon_____

(d) hydrogen _____

(e) aluminum_____

Answers: elements: c, d, and e; compounds: a and b

2

Any compound can be broken into its constituent elements by chemical means. For example, if an electric current is passed through water, the water will separate into hydrogen and oxygen. The properties of a compound may be greatly different from the properties of the elements that make it up. You probably know enough about the properties of hydrogen, oxygen, carbon, water, and carbon dioxide to compare the compound to its elements in these cases. An even more outstanding example is common table salt, or NaCl; one of the NaCl elements is a metal, and one is a poisonous gas.

Table 5.1

Number	Element	Symbol	Number	Element	Symbol
1.	hydrogen	H	56.	barium	Ba
2.	helium	He	57.	lanthanum	La
3.	lithium	Li	58.	cerium	Ce
4.	beryllium	Be	59.	praseodymium	Pr
5.	boron	B	60.	neodymium	Nd
6.	carbon	C	61.	promethium	Pm
7.	nitrogen	N	62.	samarium	Sa
8.	oxygen	O	63.	europium	Eu
9.	fluorine	F	64.	gadolinium	Gd
10.	neon	Ne	65.	terbium	Tb
11.	sodium	Na	66.	dysprosium	Dy
12.	magnesium	Mg	67.	holmium	Ho
13.	aluminum	Al	68.	erbium	Er
14.	silicon	Si	69.	thulium	Tm
15.	phosphorus	P	70.	ytterbium	Yb
16.	sulfur	S	71.	lutetium	Lu
17.	chlorine	Cl	72.	hafnium	Hf
18.	argon	Ar	73.	tantalum	Ta
19.	potassium	K	74.	tungsten	W
20.	calcium	Ca	75.	rhenium	Re
21.	scandium	Sc	76.	osmium	Os
22.	titanium	Ti	77.	iridium	Ir
23.	vanadium	V	78.	platinum	Pt
24.	chromium	Cr	79.	gold	Au
25.	manganese	Mn	80.	mercury	Hg
26.	iron	Fe	81.	thallium	Tl
27.	cobalt	Co	82.	lead	Pb
28.	nickel	Ni	83.	bismuth	Bi
29.	copper	Cu	84.	polonium	Po
30.	zinc	Zn	85.	astatine	At
31.	gallium	Ga	86.	radon	Rn
32.	germanium	Ge	87.	francium	Fr
33.	arsenic	As	88.	radium	Ra
34.	selenium	Se	89.	actinium	Ac
35.	bromine	Br	90.	thorium	Th
36.	krypton	Kr	91.	protactinium	Pa
37.	rubidium	Rb	92.	uranium	U
38.	strontium	Sr	93.	neptunium	Np
39.	yttrium	Y	94.	plutonium	Pu
40.	zirconium	Zr	95.	americium	Am
41.	niobium	Nb	96.	curium	Cm
42.	molybdenum	Mo	97.	berkelium	Bk
43.	technetium	Tc	98.	californium	Cf
44.	ruthenium	Ru	99.	einsteinium	Es
45.	rhodium	Rh	100.	fermium	Fm
46.	palladium	Pd	101.	mendelevium	Md
47.	silver	Ag	102.	nobelium	No
48.	cadmium	Cd	103.	lawrencium	Lr
49.	indium	In	104.	?*	Rf?
50.	tin	Sn	105.	?*	Ha?
51.	antimony	Sb	106.	?*	Sg?
52.	tellurium	Te	107.	?*	Ns?
53.	iodine	I	108.	?*	Hs?
54.	xenon	Xe	109.	?*	Mt?
55.	cesium	Cs	110.	?*	
			111.	?*	

*The names for these elements have not been agreed upon. The American Chemical Society has recommended the names rutherfordium, hahnium, seaborgium, nielsbohrium, hassium, and meitnerium for elements 104 through 109 respectively. The periodic table of the elements (on page 58) uses the abbreviations for these recommended names.

(a) Refer back to Table 5.1. Which elements make up table salt? _____

_____

(b) Can table salt be changed from a compound into two elements? _____

Answers: (a) sodium and chlorine;   (b) yes

ATOMS AND MOLECULES

Suppose you take an iron bar and start cutting it into smaller and smaller pieces. Even with an ideally sharp knife and a perfect microscope, you would not be able to keep cutting up the iron indefinitely. You would finally reach the smallest piece of iron that exists; you would have an **atom** of iron. An atom is the smallest part of an element that retains the properties of the element. When the atom itself is broken into parts, the resulting particles no longer retain the properties of the original element. The study of what happens when an atom is broken apart is the subject of nuclear physics.

(a) Is an atom the smallest piece of an element or of a compound? _____

(b) When you divide an atom, do you still have the same substance? _____

Answers: (a) an element;   (b) no

Elements always combine in certain fixed proportions when they form a compound. For example, it always happens that when hydrogen and oxygen combine to form water, the ratio of oxygen to hydrogen is 8 to 1 by weight. If 3 grams of hydrogen are used in the reaction, 24 grams of oxygen are used. This combination by definite, fixed proportions occurs in every chemical reaction. The easiest way to explain this **Law of Definite Proportions** is that when two elements combine, there is some smallest chunk of one element that unites with one, or two, or three, or more smallest chunks of the other element.

These smallest chunks are atoms. When water is formed from hydrogen and oxygen, two atoms of hydrogen combine with each atom of oxygen. This combination of two hydrogen atoms and one oxygen atom is called a water **molecule**. A molecule of water is written symbolically as H_2O, the subscript 2 indicating that there are two atoms of hydrogen in the molecule.

(a) In four molecules of water, how many hydrogen atoms are there? _____

(b) Describe the atomic composition of carbon dioxide (CO_2) _____ .

_____

Answers: (a) eight;   (b) one carbon atom and two oxygen atoms

5

The chemical symbol for sulfur is S, and sulfuric acid has the formula H_2SO_4. This means that a molecule of sulfuric acid contains two atoms of hydrogen, one atom of sulfur, and four atoms of oxygen. A chemical formula such as H_2O or H_2SO_4 is used to symbolize two different things: sometimes (as above) it refers to a single molecule of the compound and sometimes it refers to the compound as a whole. The meaning is normally clear from the context, so there should be no confusion.

(a) HCl is hydrochloric acid. How many hydrogen atoms would twenty molecules of this acid contain? _____

(b) How many times more oxygen atoms than hydrogen atoms does H_2SO_4 contain? _____

(c) If each oxygen atom has 16 times as much mass as each hydrogen atom, what is the ratio of the mass of oxygen in H_2SO_4 to the mass of hydrogen?

_____

Answers: (a) 20; (b) two times; (c) 32 times as much mass of oxygen (This, then, is an example of the Law of Definite Proportions.)

6

You will never see an atom, not because we can't make microscopes powerful enough, but because it is simply *impossible* due to the nature of light and the properties of atoms. (We'll have more about the nature of light later.) You undoubtedly have seen drawings of atoms that make the atom appear somewhat like the solar system, with small objects revolving around a central object. (See the figure.) Such a drawing is simply a visualization of what is called the Bohr model of the atom.

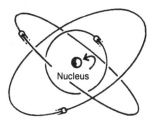

Nucleus

(a) Does an atom really look like the figure? _____

(b) Will a microscope someday enable you to see an atom? _____

Answers: (a) no; (b) no (not a microscope using visible light, anyway)

7

The Bohr model of the atom is a theoretical construct that helps us visualize the behavior of atoms in already-tested situations and to predict their behavior in situations in which they have not been observed. This widely used model is named after Niels Bohr, a Danish physicist who was a leader in the development of our concept of the atom.

The Bohr atom is made up of three particles: the proton, the electron, and the neutron. The proton has a positive electrical charge and exists in the nucleus with the electrically neutral neutrons. The negatively charged electrons revolve around the nucleus, in all directions and in all planes, as in the figure of the previous frame. (The planets of the solar system all revolve in the same direction and nearly in the same plane, so the atom is not very similar to the solar system.)

(a) If each electron contains the same amount of negative electrical charge as a proton does positive charge, what is the resultant electrical charge on an atom with two protons and two electrons? _____

(b) If an atom is normally electrically neutral, what can you say about the number of electrons and protons?

_____

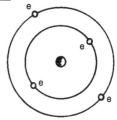

(c) Refer to the nearby figure. How many protons would this atom have? _____

Answers: (a) zero, or neutral; (b) the same number of each; (c) four (to balance the electrons)

3

Since electrons revolve around the nucleus of an atom, they form its outer "surface" and determine what other atoms their atom will combine with, and in what ratio the combination will be made. In other words, they determine the chemical properties of the element. The number of electrons an atom has revolving around its nucleus, however, is normally equal to the number of protons in the nucleus. Although the atom may momentarily lose or gain an electron,* the number of protons in the nucleus cannot be changed by ordinary means. (A nuclear reaction is required.)

(a) Which particles in the atom interact with those of another atom? _____

(b) To determine whether a particular atom is of a different element than another, what do you need to know about the atoms? _____

Answers: (a) the electrons; (b) the number of protons in their nuclei

9

Refer back to Table 5.1 and note that each element has a number to the left of its name. This number is the number of protons in the nucleus of an atom of that element. Because of the importance of this number, it is given a name, the **atomic number**.

(a) What is the atomic number of uranium? _____

(b) What element has only one proton in its nucleus? _____

(c) What does the atomic number tell us? _____

_____

*Such an atom with a fewer- or a greater-than-normal number of electrons, is called an **ion**.

(d) How many electrons does lithium have when it is neutral? _____

Answers: (a) 92; (b) hydrogen; (c) the number of protons in the nucleus of the atom; (d) three (because it has three protons)

10 THE PERIODIC TABLE

In the table on page 58, the elements are sorted by chemical properties and are listed horizontally in order by atomic number. Elements with similar properties have been placed in the same vertical column. (The properties of elements in the same column are *similar*, but not the same.) The most conspicuous example of similarity of properties is found in the last column on the right. These elements are called the noble (or inert) gases, because none of them reacts readily with another element. This table is called the **periodic table**. To familiarize yourself with the table, use it to answer these questions.

(a) Which number in each box is the atomic number? _____

(b) What is the symbol for element number 44? _____

(c) Is oxygen more like sulfur or hydrogen in its properties? _____

(d) Name two noble gases. _____

Answers: (a) the one at the top; (b) Ru (which stands for ruthenium); (c) sulfur; (d) any two from among this list: helium, neon, argon, krypton, xenon, radon

11 ELECTRON ORBITS

The figure in frame 7 showed the Bohr model with two electron orbits, but with more than one electron in each orbit. In the Bohr model, a number of electrons will normally occupy a single orbit (or level), but the maximum number that can be at any level is strictly limited (two in the first, eight in the second, and eight in the third).* For example, helium has two electrons orbiting at the first level; lithium has the third electron in a higher orbit. See the figure to answer the following questions.

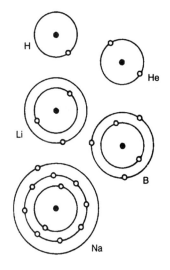

(a) Boron (B) has five electrons. How many will be in its outer orbit? _____

---

*This system of limiting the number of electrons in an orbit and *excluding* others is called the **Pauli Exclusion Principle**.

Periodic Table of the Elements

| 1 H 1.0079 | ← atomic number |
| | ← atomic weight |

| Metal |
| Metalloid |
| Nonmetal |

*Lanthanide series
**Actinide series

IA	IIA	IIIB	IVB	VB	VIB	VIIB	VIII	VIII	VIII	IB	IIB	IIIA	IVA	VA	VIA	VIIA	Noble Gases VIIIA
1 H 1.0079																	2 He 4.00260
3 Li 6.941	4 Be 9.01218											5 B 10.81	6 C 12.011	7 N 14.0067	8 O 15.9994	9 F 18.99840	10 Ne 20.179
11 Na 22.98977	12 Mg 24.305											13 Al 26.98154	14 Si 28.0855	15 P 30.97376	16 S 32.06	17 Cl 35.453	18 Ar 39.948
19 K 39.0983	20 Ca 40.08	21 Sc 44.9559	22 Ti 47.88	23 V 50.9415	24 Cr 51.996	25 Mn 54.9380	26 Fe 55.847	27 Co 58.9332	28 Ni 58.69	29 Cu 63.546	30 Zn 65.38	31 Ga 69.72	32 Ge 72.59	33 As 74.9216	34 Se 78.96	35 Br 79.904	36 Kr 83.80
37 Rb 85.4678	38 Sr 87.62	39 Y 88.9059	40 Zr 91.22	41 Nb 92.9064	42 Mo 95.94	43 Tc 98.9072	44 Ru 101.07	45 Rh 102.9055	46 Pd 106.42	47 Ag 107.868	48 Cd 112.41	49 In 114.82	50 Sn 118.69	51 Sb 121.75	52 Te 127.60	53 I 126.9045	54 Xe 131.29
55 Cs 132.9054	56 Ba 137.34	57* La 138.9055	72 Hf 178.49	73 Ta 180.9479	74 W 183.85	75 Re 186.207	76 Os 190.2	77 Ir 192.22	78 Pt 195.08	79 Au 196.9665	80 Hg 200.59	81 Tl 204.383	82 Pb 207.2	83 Bi 208.9804	84 Po (209)	85 At (210)	86 Rn (222)
87 Fr (223)	88 Ra 226.0254	89** Ac 227.0278	104 Rf (261)	105 Ha (262)	106 Sg (263)	107 Ns (262)	108 Hs (265)	109 Mt (266)	110 Discovered November 1994	111 Discovered December 1994							

*	58 Ce 140.12	59 Pr 140.9077	60 Nd 144.24	61 Pm (145)	62 Sm 150.36	63 Eu 151.96	64 Gd 157.25	65 Tb 158.9254	66 Dy 162.50	67 Ho 164.9304	68 Er 167.26	69 Tm 168.9342	70 Yb 173.04	71 Lu 174.967
**	90 Th 232.0381	91 Pa 231.0359	92 U 238.029	93 Np 237.0482	94 Pu (244)	95 Am (243)	96 Cm (247)	97 Bk (247)	98 Cf (251)	99 Es (252)	100 Fm (257)	101 Md (258)	102 No (259)	103 Lr (260)

(b) How many electrons does sodium (Na) have all together? _____

In the innermost orbit? _____

In the second orbit? _____

In the outer orbit? _____

Answers: (a) three; (b) 11, two, eight, one

12 There is a tendency for an atom to want to have its outer electron level full. This tendency can be used to explain some of the chemical behavior of atoms. Let's look at an example.

(a) Which atom has only two electrons? _____

(b) Which atom has two electrons in its first orbit and eight in its next orbit (and no other electrons)? _____

(c) Which atom has its first three orbits full (18 electrons in all)? _____

(d) Now refer to the periodic table. Which column are these in? _____
_____

Answers: (a) helium; (b) neon; (c) argon; (d) far right (the noble gases)

13 The noble gases, as you have seen, all have their outer orbits of electrons full. They are "satisfied" and have no need to share electrons with nearby atoms. They therefore do not react chemically with other atoms. The atoms of all of the elements in the column just to the left of the noble gases (7A) lack one electron from having a full outer orbit. Because of this they tend to combine with atoms having an "extra" electron in their outer orbit. It is found that atoms in the first column (1A)—lithium, sodium, potassium (K), and so on—unite with the column 7A elements in a 1-to-1 ratio to form molecules such as NaCl, KCl, and LiF.

With elements of which column would you expect column 2A elements to unite in a 1-to-1 ratio? _____

Answer: 6A (Column 2A elements have two extra electrons in their outer shells. The elements of column 6A lack two electrons from having full shells.)

The discussion of the number of electrons allowed in various orbits and the methods by which atoms combine to form molecules is not always as simple as indicated here. These simple methods work only for atoms of small atomic number.

14 | ATOMIC MASSES

The number shown in the bottom of each box in the periodic table is the relative mass of an individual atom of that element. It is called the **atomic mass** (or **atomic weight**). The unit upon which it is based is the **atomic mass unit,** or amu.

(a) What is the mass of a hydrogen atom? _____ amu

(b) What is the mass of an oxygen atom? _____

(c) Is the atomic mass greater or less than the atomic number? _____

Answers: (a) 1.0079; (b) 15.9994 amu; (c) greater

15 | AVOGADRO'S NUMBER

The atomic mass of hydrogen is about 1. The atomic mass of carbon is 12. Thus, one atom of carbon has a mass 12 times as great as one atom of hydrogen. And 359 atoms of carbon have a total mass 12 times the mass of that many hydrogen atoms. The number "359" is not special though; I just picked it randomly. In fact, $6.02 \cdot 10^{23}$ atoms of carbon have a total mass 12 times the mass of that many hydrogen atoms.* This last number, $6.02 \cdot 10^{23}$, is a special one, because that many hydrogen atoms have a mass of 1 gram. It follows that $6.02 \cdot 10^{23}$ carbon atoms have a mass of 12 grams, then. And since the atomic mass of oxygen is 16, $6.02 \cdot 10^{23}$ oxygen atoms have a mass of 16 grams. This number of atoms is called Avogadro's number, and the mass of an element having that many atoms is called 1 **mole** of the substance. A mole of any substance is defined as that amount of the substance that contains Avogadro's number of elemental particles of the substance. Look up the atomic mass of chlorine in the periodic table.

(a) One mole of chlorine contains how many grams? _____

(b) How many atoms of chlorine are found in 1 mole of chlorine? _____

Answers: (a) 35.5; (b) $6.02 \cdot 10^{23}$

16

Now we move from atoms to molecules. The mass of one water molecule is 18 times the mass of one hydrogen atom, since it contains two hydrogen atoms and one oxygen atom, the oxygen's atomic mass being 16. The mole of water is therefore 18 grams.

---

*See Appendix I for an explanation of powers-of-ten notation.

(a) Use the periodic chart to calculate the mass of a mole of sulfuric acid (H_2SO_4).

_____

(b) How many molecules will a mole of sulfuric acid contain? _____

Answers: (a) 98.1 grams; (b) $6.02 \cdot 10^{23}$

17 PROBLEMS (OPTIONAL)

Now try a few problems.

(a) How many molecules are in 44 grams of CO_2 (carbon dioxide)? _____

(b) How many molecules are in 11 grams of CO_2? _____

(c) What is the mass (in grams) of one molecule of CO_2? (Hint: There are 44 g in $6.02 \cdot 10^{23}$ molecules.) _____

Answers: (a) $6.02 \cdot 10^{23}$; (b) $1.51 \cdot 10^{23}$ (We have $\frac{1}{4}$ mole, since $\frac{11}{44} = \frac{1}{4}$.); (c) $7.31 \cdot 10^{-23}$ grams.

Solution: $\dfrac{44 \text{ g}}{6.02 \cdot 10^{23} \text{ molecules}} = 7.31 \cdot 10^{-23}$ g/molecule

SELF-TEST

Refer back to the table of elements and the periodic table at any time.

1. Identify the substances below as elements or compounds

   (a) berkelium _____   (b) ammonia _____

   (c) MgO _____   (d) tin _____

   (e) Hf _____

2. What do we call the smallest unit an element can be broken into and still have the same properties? _____

3. What do we call the smallest unit of a compound that retains the properties of the compound? _____

4. In the figure on the next page, mark the location of the electron, the proton, and the neutron. Mark the part of the atom that is positively charged and the part that is negatively charged.

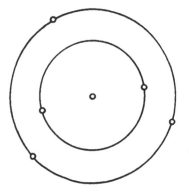

5. What does the atomic number tell us about an atom? _____

_____

6. Use the periodic table to find the element with atomic number 27. _____
   With atomic mass 238. _____

7. Explain, on the basis of electron orbits, why gases such as helium do not enter
   into chemical reactions. _____

8. The magnesium atom has two electrons in its outer orbit. Name two other elements whose atoms have this characteristic. _____

9. What is meant by the atomic mass of an element? _____

   _____

10. Use the periodic table to find the atomic number and atomic mass of sodium
    (Na). _____

11. *Optional.* How many molecules are contained in 2.8 grams of carbon monox-
    ide (CO)? _____

12. *Optional.* What mass of gold (Au) is required to have 10^{25} gold atoms?

    _____

4.

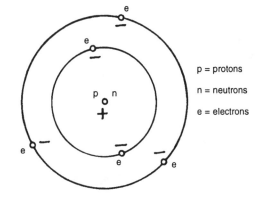

p = protons

n = neutrons

e = electrons

(frame 7)

5. the number of protons (frame 9)

6. Co (cobalt); U (uranium) (frame 10)

7. Their outermost electron orbit is full. (frames 11–13)

8. Be (beryllium), Ca (calcium), Sr (strontium), Ba (barium), Ra (radium) (frames 11–13)

9. the relative mass of the atom, compared to other atoms (frame 14)

10. 11, 23.0 (frames 10, 14)

11. $6.02 \cdot 10^{22}$ molecules (From the periodic table we learn that carbon monoxide has a molecular mass of 28, or 12 + 16. Thus, we have 0.1 mole of the compound, 0.1 times Avogadro's number yields the answer.) (frame 16)

12. 3270 grams
Solution:

10^{25} molecules is $\dfrac{10^{25}}{6.02 \cdot 10^{23}}$ or 16.6 mole

One mole of gold (atomic mass 197) has a mass of 197 grams, so we need 16.6 times this, or 3270 grams. (frame 17)

6 Solids

Prerequisite: Chapter 5

We have daily contact with matter in three states: solid, liquid, and gas. This chapter is concerned with solids, while the next deals with liquids and gases. The solids we encounter are as different as apples and suede hats, or ice cubes and horseshoes. But there are similarities, too, and science progresses largely by looking for similarities among objects.

OBJECTIVES

After completing this chapter, you will be able to

- describe the arrangement of atoms or molecules in a solid;
- differentiate between mass density and weight density;
- calculate an object's density, weight (or mass), or volume, given the other two quantities;
- given the specific gravity of a material, calculate its density;
- calculate the pressure exerted, given the force and area;
- describe a perfectly elastic substance;
- use Hooke's law to determine force or deformation;
- state the conditions under which Hooke's law applies.

1 ATOMS IN A SOLID

When matter is in a solid state, its atoms or molecules are arranged in definite, fixed patterns.* The atoms are held in their positions by electrical forces between

---

*In a different type of solid, not even called a solid in some books, the atoms and molecules have a random arrangement. Glass is of this type, called an **amorphous solid** or a **supercooled liquid**. Most of us would rather not call glass a liquid, but what about margarine—is it solid or liquid? (It is amorphous.) In this chapter, we will assume that "solid" actually means "crystalline solid." (The word "crystalline" tells us definitely that the solid has a regular molecular arrangement.)

each atom and its neighbors. These forces give the solid its rigidity. The result of the regular arrangement of molecules is evident in the shape of snowflakes, which—although they are very different in detail—all have a six-sided symmetry. Other arrangements of molecules give characteristic shapes for other crystalline structures.

(a) Every grain of table salt has a characteristic cubic shape. What property of solids causes this? _____

(b) What could you predict about the arrangement of molecules in something that is *not* a solid? _____

Answers: (a) a fixed pattern of molecular arrangement; (b) random or irregular

DENSITY

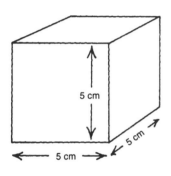

Solids differ in many ways, including color, texture, and elasticity. Another important distinguishing property of solids is density. Density is defined in two different ways, and to avoid confusion we will call one—the more fundamental one—"mass density" and the other "weight density."

Mass density is defined as the ratio of the mass of a substance to its volume. For example, suppose we have a steel cube 5 cm by 5 cm by 5 cm. Its mass (measured on a balance) is 975 grams.

(a) What is its volume? _____

(b) What is the mass density of the steel? _____ g/cm^3

Answers: (a) 125 cm^3; (b) 7.8 (which is $\dfrac{975 \text{ g}}{125 \text{ cm}^3}$)

A larger piece of steel on which you make the same measurements and do the same calculations will have the same value for its mass density, since density is characteristic of the material and does not depend upon how much material is present.

The density of aluminum is 2.7 g/cm^3. Suppose you have 12 cm^3 of aluminum.

(a) What is its mass? _____

(b) What is the mass density of a piece of aluminum 24 cm^3 in volume? _____

Answers:
(a) 32.4 g. Solution: density = mass/volume
$\qquad\qquad\qquad$ mass = density·volume
$\qquad\qquad\qquad\qquad$ = 2.7 g/cm^3·12 cm^3
$\qquad\qquad\qquad\qquad$ = 32.4 g
(b) 2.7 g/cm^3

4

If mass is measured in kilograms and volume in cubic meters, density has units of kg/m^3. One g/cm^3 is equivalent to 1000 kg/m^3. Table 6.1 shows the densities of some common substances.

Table 6.1

Substance	Mass density (g/cm³)	Weight density (lb/ft³)
steel	7.8	488
aluminum	2.7	168
bone	1.8–2.0	112–125
mercury	13.6	849
ice	0.92	57.4
wood	0.35–.85	22–53
lead	11.3	705
water	1.00	62.4

Weight density is defined as the weight of a substance divided by its volume and is normally used in the British system of units. The unit of weight density in this system is lb/ft^3 (or $lb/in.^3$).

(a) Steel has a weight density of 488 lb/ft^3. How much does a cubic foot of steel weigh? _____

(b) What is the weight of a block of lead that measures 6 inches by 6 inches by 1 foot? (See Table 6.1.) _____

(c) What is the volume (in m^3) of 5000 kilograms of ice? _____

Answers:
(a) 488 lb
(b) 176 lb. Solution: volume = 0.5 ft·0.5 ft·1 ft = 0.25 ft^3
        weight = density·volume
              = 705 lb/ft^3·25 ft^3
              = 176.25 lb/ft^3
(c) 5.43 m^3. Solution: First, find the density in units of kg/m^3:

$$0.92 \text{ g/cm}^3 \cdot \frac{1000 \text{ kg/m}^3}{1 \text{ g/cm}^3} = 920 \text{ kg/m}^3$$

density = mass/volume
volume = mass/density
$$= \frac{5000 \text{ kg}}{920 \text{ kg/m}^3}$$
$$= 5.43 \text{ m}^3$$

5 SPECIFIC GRAVITY

The **specific gravity** of a substance is often listed in tables instead of either mass density or weight density. Specific gravity is defined as the ratio of the density of the substance in question to the density of water. The mass density of water* is 1.00 g/cm^3. Thus, the specific gravity of lead (density 11.3 g/cm^3) is 11.3 (because it is (11.3 g/cm^3)/(1 g/cm^3)). In units gm/cm^3, the specific gravity of a substance is always numerically equal to its mass density. Notice, however, that specific gravity has no units, since it is simply a ratio of two densities.

The specific gravity of rock is about 5. (Since rock is of different types, this is approximate.) What is the weight of a cubic yard of rock? (You can either work this out yourself or go through the steps below.)

(a) What is the density of rock in lb/ft^3? (Use Table 6.1 to find the density of water.) _____

(b) How many cubic feet are in a cubic yard? _____

(c) What is the weight of a cubic yard of rock? _____

Answers: (a) 310 lb/ft^3 (Since the weight density of water is 62.4 lb/ft^3, the density of rock is 5·62.4 lb/ft^3. We round the answer because the density is not well known, so the answer can't be.); (b) 27 ft^3 (3 ft·3 ft·3 ft); (c) 8400 lb (310 lb/ft^3·27 ft^3)

6 PRESSURE

The total amount of force on an object is not always the critical factor. For example, you might stack two or three books on your head without feeling pain, but if you put a tiny pebble between the bottom book and your skull, the sensation could be very unpleasant. The important factor here is force divided by the area over which the force is applied, or the **pressure**.

Let's calculate the pressure exerted on the floor by a 200-lb man standing on both feet. If the contact area of each shoe is 3 in. by 10 in., the area of contact is 30 in.^{2} for each shoe. What is the pressure exerted on the floor? (Include units in your answer.)

Answer: 3.33 lb/in.^{2} (200 lb/60 in.^{2})

---

*Notice that density is defined for liquids as well as solids—in fact, water is the reference point for specific gravity.

7

In SI units, the unit of force is the newton. (See Chapter 2 for further details, but all that is needed here is the unit's name. You might be interested to know that a newton is about a quarter pound.) The 200-lb man mentioned above weighs about 800 N, and one shoe's dimensions are about 8 cm by 25 cm.

(a) Calculate the pressure he exerts on the floor (in N/cm²) when he stands on one foot. _____

(b) Express your answer to part (a) in N/m². _____

Answers: (a) 4 N/cm²; (b) 40,000 N/m² (There are 100², or 10,000 cm², in 1 m².)

From this example, you see that units of N/m² result in large numbers in many common examples. This is because a square meter is a large area.

8

In the 1950s, when women wore spike heels with areas of about $\frac{1}{4}$ cm² ($\frac{1}{2}$ cm by $\frac{1}{2}$ cm), aircraft companies were concerned because such heels could exert enough pressure to punch through the floor of planes. What pressure would a 450-N (110-lb) woman exert on the floor when she puts her entire weight on one heel? Express your answer in N/cm². _____

Answer: 1800 N/cm² (Compare this to the 4 N/cm² exerted by a man standing on one foot.)

9 ELASTICITY AND HOOKE'S LAW

Many solids can be deformed but will return to their original shape when the cause of the deformation is removed. The elasticity of a substance is a measure of how well it returns to its original shape. According to this definition, we find that steel must be classified as more elastic than rubber. Although steel is not easily deformed, it returns to its original shape more readily than does rubber.

(a) Would you classify such materials as dough and clay as elastic or inelastic? Why? _____

(b) Which type of substance returns exactly to its original form, perfectly elastic or perfectly inelastic? _____

Answers: (a) inelastic, because they don't return to original shape; (b) perfectly elastic

Everyday substances lie somewhere between these two extremes. (You may recall the discussion of elastic and inelastic collisions in Chapter 3. The use of the term is similar here, for if an object returns completely to its original shape, no energy is lost in deforming it.)

10 Hooke's law states that the amount of deformation produced by a force is proportional to the amount of force. Only a perfectly elastic object obeys Hooke's law perfectly. A steel spring is close enough to perfectly elastic that we can consider it so, as long as it is not strectched too far. Suppose you put a 1-pound weight on such a spring and the spring stretches 0.75 in. A 2-pound weight will cause a deformation of 1.5 in. In equation form we can write:

$$F = k \cdot x$$

where **F** is the force applied, k is a constant (called the spring constant), and **x** is the deformation. Since a force of 1 lb produces a 0.75-in deformation, the spring constant k for the spring above is 1.33 lb/inch.

(a) How much deformation will a 3-lb force cause? _____

(b) How much force is required to stretch the spring 3 in.? _____

Answers: (a) 2.25 inches;   (b) 4 lb

11 The spring constant tells us how "stiff" a spring is. Different springs have different spring constants.

(a) How much deformation is produced by a force of 2 N if the spring constant is 2 N/cm? _____

(b) Will a spring with a constant of 5 N/cm require more or less force to stretch it a given distance than a spring with a constant of 8 N/cm? _____

Answers: (a) 1 cm;   (b) less

SELF-TEST

Refer to the table of densities (page 66) at any time.

1. Crystals of common table salt tend to have 90° angles between their flat sides. What causes this? _____

_____

2. A gallon of water weighs 8 lb. What is its volume in cubic feet? _____

3. A certain block of wood has dimensions of 4 cm by 8 cm by 40 cm. If it has a mass of 800 grams, what is its density in g/cm^3? _____

4. Suppose you have 2 cubic feet of each of the materials in Table 6.1. Which will weigh the least? _____   The most? _____

5. How much would the mass of 5 cubic meters of ice be? _____

6. What is the specific gravity of a material that has a density of 2500 kg/m³?

   _____

7. Why is the specific gravity of a substance equal to its density in g/cm³?

   _____

   _____

8. Suppose the entire 3-lb weight of some books rests on your head with a wooden cube between. If the cube is $\frac{1}{4}$ inch on a side, what pressure is exerted on your head? _____

9. Does Hooke's Law apply to all objects? Why or why not? _____

   _____

10. If a spring has a spring constant of 3 lb/inch, how far would it stretch if 10 lb were hung from its end? _____

1. The regular arrangement of atoms within the crystal causes the regularity of shape. (frame 1)

2. 0.128 ft³ (8 lb/62.4 lb/ft³) (frames 2–4)

3. 0.63 g/cm³. Solution:
$$\text{volume} = 4 \text{ cm} \cdot 8 \text{ cm} \cdot 40 \text{ cm} = 1280 \text{ cm}^3$$
$$\text{density} = 800 \text{ g}/1280 \text{ cm}^3 = 0.63 \text{ g/cm}^3 \text{ (frame 4)}$$

4. wood, mercury (frame 4)

5. 4600 kg. Solution: density $= 0.92$ g/cm³
$$5 \text{ m}^3 \cdot \frac{100^3 \text{ cm}^3}{1 \text{ m}^3} = 5,000,000 \text{ cm}^3$$
$$\text{weight} = \text{density} \cdot \text{volume}$$
$$= 0.92 \text{ g/cm}^3 \cdot 5,000,000 \text{ cm}^3$$
$$= 4,600,000 \text{ g, or } 4600 \text{ kg (frame 4)}$$

6. 2.5. Solution: The density of water is 1000 kg/m³.
$$\text{specific gravity} = \frac{\text{density of material}}{\text{density of water}}$$
$$= \frac{2500 \text{ kg/m}^3}{1000 \text{ kg/m}^3}$$
$$= 2.5 \quad \text{(frame 5)}$$

7. Specific gravity is defined as the ratio of the density of the substance to the density of water. Since the density of water is 1.0 g/cm³, the specific gravity is numerically equal to the density. (frame 5)

8. 48 lb/in² (The area of one face of the cube is $\frac{1}{4}$ in. $\cdot \frac{1}{4}$ in. or $\frac{1}{16}$ in.². The pressure is thus 3 lb $\div \frac{1}{16}$ in.².) (frames 6–8)

9. No. It applies exactly only to materials that are perfectly elastic. Some materials are very inelastic, and Hooke's Law does not apply to them at all. (frames 9, 10)

10. 3.33 inches (It requires 3 lb to stretch an inch. Ten pounds would stretch it 10/3 times that far.) (frames 10, 11)

<u>7</u> Liquids and Gases

Prerequisites: Chapters 5 and 6
Of the three states of matter—solid, liquid, and gas—both liquids and gases are considered fluids. Although in everyday language we often consider the word fluid to apply only to liquids, in physics gases are considered fluids also, since gases and liquids behave similarly in many situations.

OBJECTIVES

After completing this chapter, you will be able to

- compare liquids and gases to solids in terms of molecular attraction and organization;
- compare liquids to solids in terms of elasticity;
- calculate pressure at a given depth in a given liquid;
- use Archimedes' principle to find the buoyant force on a given object in a given fluid;
- calculate the depth at which a given object will float in a specified fluid;
- use Pascal's principle to demonstrate how pressure is transmitted in a liquid;
- compare diffusion rates in liquids and gases and use kinetic theory to explain the difference;
- calculate the total force on a given object due to atmospheric pressure;
- relate atmospheric pressure to "inches of mercury";
- relate the pressure of a confined gas to the number of molecules and the average speed of the molecules.

1 MOLECULES IN A LIQUID

A liquid is distinguished from a solid in that a liquid flows to fill the bottom of a container until the liquid has a level surface. In a solid, molecules are held to a definite position in a regular pattern (although they may vibrate about a central position), while in a liquid the molecules are free to move about one another. A force is indeed exerted between neighboring molecules of a liquid, as indicated by the fact that molecules do not easily leave the surface. Nor does a liquid contract easily. If enough pressure is exerted on a liquid to cause it to contract, its original volume is restored when the force is removed.

(a) Are molecules arranged in a liquid regularly or randomly? _____

(b) Are molecules in a liquid more or less tightly bound together than in a solid? _____

(c) In general, are liquids elastic or inelastic? _____

Answers: (a) randomly; (b) less tightly; (c) elastic (see Chapter 6, frame 9)

2 PRESSURE IN A LIQUID

Most of us experience the feeling of increased pressure on our ears when we dive deep below the surface of a swimming pool. Pressure in a liquid increases with depth below the surface and can be calculated by the following simple formula:

$$\text{pressure} = \text{depth} \cdot \text{weight density}$$

(a) What is the water pressure at a depth of 10 feet below the surface of water? (The weight density of water is 62.4 lb/ft^3.) _____

(b) What is this pressure in units of lb/in.^{2}? _____

Answers: (a) 624 lb/ft^2 (10 ft·62.4 lb/ft^3); (b) 4.3 lb/in.^{2} (There are 144 square inches in a square foot. So, 624 lb/ft^2 is equal to 624/144, or 4.3 lb/in.^{2}.)

3 Now assume the area of the eardrum to be $\frac{1}{16}$ in.^{2} (which is $\frac{1}{4}$ in. by $\frac{1}{4}$ in.), and calculate the force exerted on it by the water when you are 10 feet under.

Answer: 0.27 lb (4.3 lb/in.^{2}·$\frac{1}{16}$ in.^{2})
(The fact that we can feel this small force indicates that the ear is very sensitive.)

4 In SI units, density is expressed as kg/m^3. This is a mass density, so we cannot simply multiply density times depth to obtain pressure. First we must calculate the weight density. Recall that 1 kilogram has a weight of 9.8 newtons,* and that in general we multiply mass in kilograms by 9.8 to obtain weight in newtons. Therefore, the weight density of water (1000 kg/m³) is 9800 N/m³. Now we'll calculate the pressure at a depth of 5 meters in alcohol (which has a specific gravity of 0.80).

(a) What is the mass density of alcohol? _____

(b) What is the weight density of alcohol? _____

(c) What is the pressure 5 meters deep in alcohol? _____

(d) One square meter is equal to 10,000 square centimeters. What is the pressure in N/cm²? _____

Answers: (a) 800 kg/m³ (see Chapter 6, frame 2); (b) 7840 N/m³; (c) 39,200 N/m²; (d) 3.92 N/cm²

5 BUOYANCY

Consider a cubic foot of water in a swimming pool full of water. (To do this you might think of the cubic foot being enclosed by thin tissue paper.) The water weighs 62.4 pounds. Since it is held there in the water of the pool, the water around it must be exerting an *upward* force of 62.4 pounds on it. If somehow the cubic foot of water were removed and an equal volume of something else put in its place, that something else would also feel an upward force of 62.4 pounds exerted by the water around it. We call this upward force the *buoyant force*, and the example above illustrates the general rule for determining the magnitude of the buoyant force:

> An object in a fluid is buoyed up by a force
> equal to the weight of the fluid displaced.

This principle, called **Archimedes' principle,** is important enough to deserve some examples.

Suppose a cubic foot of steel (weight: 480 lb) is below the surface of water.

(a) What is the buoyant force exerted on it by the water? _____

(b) What, then, will be the apparent weight of the steel? (In other words, how much force would you have to exert to lift the steel under water?)_____

Answers: (a) 62.4 lb; (b) 418 lb (480 lb – 62 lb)

---

*If you have not studied Chapter 2, you need only know that the newton (N) is the unit of force in SI—International System—units. See frames 1, 2, 5, 7, and 8 of Chapter 2 for more details.

(Note that although pressure is greater at a greater depth, the buoyant force does not depend upon the depth. You did not have to consider how deep the steel was, but simply that it was below the surface.)

6
A cubic foot of pine wood weighs about 30 pounds. If it were pushed underwater, the buoyant force on it would also be 62.4 lb. But since the buoyant force is greater than its weight, pine will not stay underwater unless held there. Let's calculate how much of its volume will remain under water when it floats. To do this, we reason that it will sink in the water until it is buoyed up by a force that is equal to its weight.

(a) What weight of water will the 30-lb piece of wood have to displace in order to float? _____

(b) What is the volume of this weight of water? _____

(c) How much of the wood will be below water level as the block of wood floats?

_____

Answers: (a) 30 lb; (b) 0.48 ft^3 (30 lb/62.4 lb/ft^3); (c) 0.48 ft^3

7
In general, one can determine what fraction of a floating object will be under a liquid by dividing the density of the object by the density of the liquid.

(a) What is the density of the wood in the last frame? _____

(b) What is the density divided by the density of water? _____

(c) Suppose a piece of oak wood floats with 62 percent of its volume underwater. What is the density of the wood? _____

Answers: (a) 30 lb/ft^3 (The cubic foot weighed 30 lb.); (b) 0.48 (Since the wood had a volume of 1 ft^3, this corresponds to answer (b) of the last frame.); (c) 38.7 lb/ft^3 (or 620 kg/m^3) (This is 0.62 times the density of water.)

8
As in calculating pressure, we must again use the weight density rather than the mass density in calculations involving buoyancy. Let us determine the apparent weight in water of a 5-cm^3 piece of aluminum—about pencil size. (The specific gravity of aluminum is 2.7.)

(a) What is the weight density of aluminum in N/m^3? _____

(b) What is the weight density in N/cm^3? _____

(c) How much does the piece of aluminum weigh? _____

(d) What is the weight density of water in SI units? _____

(e) Convert this to N/cm³. _____

(f) What is the buoyant force on 5 cm³ of aluminum? _____

(g) What is the apparent weight of the aluminum in water? _____

Answers: (a) 26,500 N/m³ (2,700 kg/m³·9.8 N/kg); (b) 0.0265 N/cm³ (There are 10⁶ cm³ in a m³.); (c) 0.133 N (0.0265 N/cm³·5 cm³); (d) 9800 N/m³; (e) 0.0098 N/cm³; (f) 0.049 N (the weight of 5 cm³ of water); (g) 0.084 N (0.133 N – 0.049 N)

9 PASCAL'S PRINCIPLE

When additional pressure is put on a confined liquid, the pressure is transmitted equally to all parts of the liquid. This principle—**Pascal's principle**—is best understood by looking at an example. The figure illustrates a water container with two openings, each sealed by a movable piston. The openings, and therefore the pistons, have different areas, however—the left one being 10 cm² and the right one 30 cm². The liquid just below the pistons is at the same height, so the pressure is the same below each. (Remember that pressure depends upon depth and liquid density.)

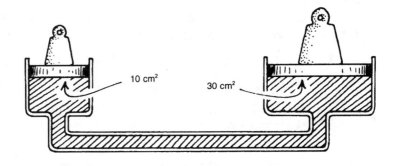

Now suppose we put an 8-newton weight on the left piston. What weight must be put on the other piston to balance this weight? To solve this, we must calculate the extra pressure exerted on the liquid due to the weight at left. This is (8 N/10 cm²), or 0.8 N/cm². Now, according to Pascal's principle, this pressure is exerted at all parts of the liquid, including on the piston at right. But the area of this piston is 30 cm², and so the force exerted on it by the extra water pressure is 0.8 N/cm²·30 cm², or 24 newtons. Thus, a 24-newton weight would be required on the right to balance the 8-newton weight on the left.

Notice that the ratio of weights is the same as the ratio of areas. That is, the piston at right has three times the area, and it requires three times as much force to balance the one on the left.

A hydraulic jack uses two pistons connected basically as shown in the figure. It can be used to lift very heavy objects because its pistons have greatly different areas.

(a) Suppose we placed a 50-kg mass on the left piston. How much mass would be needed on the right to balance it? _____

(b) In using a hydraulic jack, on which piston (smaller or larger) would a person exert his or her force? _____

Answers: (a) 150 kg;   (b) smaller

10   MOLECULES IN GASES

Most of us are familiar with the fact that when a liquid changes to a gas—as when water is boiled—the gas takes up much more room than the liquid. This happens because the molecules of the gas are much farther apart than the molecules of the liquid. In fact, the molecules of the gas are not bound together by forces at all, but fly about freely in space. They exert forces on each other only when they collide. When two molecules do collide, they bounce from each other and continue freely until the next collision.

This model of a gas, called the kinetic theory, is consistent with a number of everyday observations concerning gases. For example:

(a) Which can be compressed more easily, liquids or gases? _____

(b) Which do you expect to be less dense, liquids or gases? _____

Answers: (a) gases (because there is space between the molecules of a gas);   (b) gases (because the empty space contributes nothing to the weight)

11   DIFFUSION THROUGH LIQUIDS AND GASES

Diffusion is the intermixing of molecules of substances due to random motion of the molecules. The molecules of one liquid diffuse rather slowly through another liquid. Suppose you carefully put a drop of food coloring in the bottom of a glass of water which has been sitting long enough that the water is not swirling about. (You could do this with an eyedropper.) If you do, you will find that the color gradually spreads out through the water. It will be hours—or even days—before the color is entirely diffused through all of the water, however.

Now we'll consider one gas diffusing through another. If you spill some ammonia at one end of a room, before long you can smell the ammonia halfway across the room, and a little later the smell will have filled the entire room. This smell is simply the result of molecules of ammonia entering your nose. In a matter of minutes the ammonia molecules move as far as 10 or 20 feet.

(a) Which occurs faster, diffusion of a liquid through a liquid or a gas through a gas? _____

(b) What actually happens in diffusion? _____

_____

(c) How does the molecular spacing in a gas contribute to diffusion through a gas being faster than through a liquid? _____

_____

Answers: (a) gas through a gas; (b) The molecules of one substance move among those of another. (c) There is space between gas molecules for other molecules to move through. This is not so for a liquid; in a liquid the molecules must jostle around one another.

---

12 PRESSURE IN GASES

Gases are more easily compressed than liquids. But if there is so much empty space between molecules, why is any effort at all required to compress them? Why do they exert pressure? Let's consider blind bumblebees flying around in a boxcar. Such bees (if they are too ignorant to land, anyway) would constantly be bumping into the walls of the car. Each collision with a wall would result in a very small push against the wall, but if enough bumblebees were present, we can imagine not only that the overall force against the wall would be significant, but that the collisions would be so frequent that the individual bumps could not be distinguished. The effect would be that of a constant force. This is what happens in a gas. Molecules, of course, are both unimaginably small and unimaginably numerous, and their frequent collisions with the walls of their container produce a constant force on walls or boundaries.

Just as with liquids, it is usually more convenient to speak of the pressure (force/area) exerted by a gas rather than the total force exerted.

(a) Suppose the number of gas molecules in an enclosed container is increased. Would you expect the pressure due to the gas to increase or decrease?

_____

(b) In Chapter 8 we will see that raising the temperature of a gas increases the average speed of the molecules. Suppose the temperature of a confined gas is increased. How would this affect the pressure? _____

Answers: (a) increase (because more molecules would be hitting the walls); (b) The pressure would increase (for two reasons: (1) each collision with a wall would exert more force on the wall, and (2) the molecules would hit the walls more frequently).

13 The gas pressure we are most familiar with is that of the atmosphere. Because of the tremendous amount of air above us, the atmosphere exerts a force of about

14.7 pounds on every square inch of surface at sea level. Hold your hand horizontally in front of you, palm up. The area of the palm of your hand is about 20 square inches.

(a) How much force is the atmosphere exerting downward on your palm?

_____

(b) But the air under your hand also exerts a pressure upward. How much force is exerted on the underside of your hand? _____

Answers: (a) 294 lb (20 in.^{2}·14.7 lb/in.^{2});   (b) 294 lb
(The hand is not crushed by these two forces because we grow up with atmospheric pressure on us, and equal pressures are exerted on our flesh from the inside by the liquids and solids within our bodies.)

14 In metric units, standard atmospheric pressure is 101,000 N/m^2, or 1.01·10^5 N/m^2.

(a) How many newtons of force are exerted by the atmosphere on a desktop that has dimensions 1.5 m by 1.0 m? _____

(b) Why does the atmosphere not crush the desk? _____

_____

Answers: (a) 152,000 N;   (b) The atmosphere exerts a force upward on the underside of the desktop.

15 The pressure in a gas (as in a liquid) increases as one gets deeper and deeper in the gas. Recall that in a liquid, as one doubles his depth below the surface he doubles the pressure. The case is not quite so simple for atmospheric pressure, however. (After all, what is meant by the "surface" of the atmosphere?) The density of a liquid is the same throughout; the atmosphere, however, is less dense at greater height. Thus, the weight of atmospheric gases above one's head does not vary in any direct way with the person's height above sea level. The change in air pressure as one changes height is experienced, however, when we drive quickly up or down a long hill, or better yet when we ascend or descend in an airplane. We feel the same kind of discomfort in our ears as when we dive too deeply in a swimming pool.

The atmospheric pressure stated in an earlier frame, 14.7 lb/in.^{2}, or 1.01·10^5 N/m^2, is the average atmospheric pressure at sea level.

(a) Would you expect atmospheric pressure to be greater or less on a Colorado mountain top than at sea level? _____

(b) Would you expect the air to be more or less dense on that mountain top?

_____

Answers: (a) less;   (b) less

16 THE BAROMETER

A barometer is a device used to measure atmospheric pressure. The most basic type of barometer is shown in the figure. A long tube is filled with mercury and inverted in a pool of mercury. The mercury does not stay in the top of the tube but falls, leaving a vacuum at the top. The mercury level will fall to a height of about 30 inches (about 76 cm) above the mercury pool when the apparatus is at sea level no matter how much longer than 30 inches the tube is made.

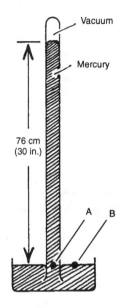

The reason that the mercury stays up in the tube is that the atmosphere, pushing down on the surface of the mercury pool, pushes it up. And the atmospheric pressure is able to hold the mercury up to a height such that the pressure at point A due to the mercury above is equal to atmospheric pressure. (The pressure at A must equal the pressure at B because the points are at the same level.) Let's calculate the pressure caused by 76 cm of mercury. The density of mercury is 13,600 kg/m^3, so that its weight density is 133,000 N/m^3.

(a) What is the formula for pressure at a depth in a liquid? _____

_____

(b) How much is the pressure at a depth of 76 cm below the level of mercury?

_____

(c) Would a reading of 74 centimeters of mercury indicate more or less pressure?

_____

Answers: (a) pressure = depth·density; (b) $1.01 \cdot 10^5$ N/m^2 (133,000 N/m^3·0.76 m); (c) less

17

Rather than stating the pressure of the atmosphere in pounds per square inch, we normally speak of the pressure being so many inches, or so many centimeters, of mercury. To be absolutely correct, we should not say that the pressure today is 29.8 inches, but that it is equal to the pressure exerted at a depth of 29.8 inches in mercury. (Weather reports on TV are limited as to time available, so the weatherperson takes a shortcut!)

Compact barometers have an airtight container with a flexible end. As atmospheric pressure increases, this flexible end is pushed inward and a lever connected to it moves a pointer on a dial on the face of the barometer. Because of the common use of the mercury barometer, such dials are normally calibrated in inches or centimeters of mercury even though they contain no mercury.

(a) What is the equivalent of 30 inches of mercury stated in mm Hg (millimeters of mercury)? (Note: 1 inch = 25.4 mm) _____

(b) When the pressure is 15 lb/in.^{2}, what would a mercury barometer read in inches? (Note: The density of mercury = 0.49 lb/in.^{3}.) _____

Answers: (a) 762 mm Hg
(b) 30.6 in. Solution: pressure = depth·density
depth = pressure/density

$$= \frac{15 \text{ lb/in.}^2}{0.49 \text{ lb/in.}^3}$$

= 30.6 in.

18 BUOYANCY IN GASES

Archimedes' principle states that an object in a fluid is buoyed up with a force equal to the weight of fluid displaced. We notice the effect of buoyant force in liquids more than in gases because the density of liquids is so much greater. The weight of liquid displaced is much greater than the weight of gas displaced by a given object. The density of air at sea level is about 0.075 lb/ft^3, and the volume of a 125-lb person is about 2 cubic feet.

(a) What is the buoyant force on this person due to the atmosphere? _____

(b) What would be the buoyant force on this person in water? _____

Answers: (a) 0.15 lb or 2.4 ounces (0.075 lb/ft^3·2 ft^3)   (b) 125 lb (62.4 lb/ft^3·2 ft^3). (The fact that this is also the weight of the person confirms the observation that people are just between floating and sinking when in water.)

19

The most visible example of the buoyant force of air is the case of helium balloons and blimps. A large blimp may have a volume of a few million cubic feet. At 0.075 lb/ft^3, a million cubic feet displaces 75,000 pounds. In order for a million-cubic foot airship to be lifted by buoyant force, all that is required is that the entire ship—structure and enclosed gas—weigh less than 75,000 pounds. The use of a very low density gas allows this to be achieved. The least dense gas is hydrogen, but it is very explosive, so helium is used. The Hindenburg was a 7 million-cubic foot airship containing hydrogen; it exploded and burned in 1937.

(a) What is the most the Hindenburg could have weighed and still flown?

_____

(b) Why do hot-air balloons rise? _____

Answers: (a) 525,000 lb;   (b) Hot air is less dense than cold air.

20 SOLIDS, LIQUIDS, AND GASES REVIEWED

Over the last two chapters we have studied the three states of matter and have seen similarities as well as differences. Of the three states—solids, liquids, and gases—which

(a) are least dense (in general)? _____

(b) have most rigid molecular bonding? _____

(c) are most compressible? _____

(d) permit diffusion? _____

Answers: (a) gases; (b) solids; (c) gases; (d) liquids and gases (Solids do diffuse through solids, but centuries must pass before the effect can be seen, and even then the effect is very slight.)

SELF-TEST

1. What is the difference between a liquid and a solid as to how the molecules are bound? _____

_____

_____

2. What is the relationship between pressure, depth, and density in a liquid?

_____

3. Calculate the pressure at a depth of 20 feet below the surface of water (density 62.4 lb/ft^3). _____

4. What is the pressure (in SI units) at a depth of 3 meters in alcohol (specific gravity 0.80)? _____

5. A basketball has a volume of about 0.52 cubic feet. What is the buoyant force on it if it is pushed underwater? _____

6. A certain cornerstone for a building is 1 ft by 1 ft by 2 ft, and weighs 425 lb. What will be its apparent weight underwater? _____

7. Mercury has a specific gravity of 13.6. How much buoyant force is exerted on a 100-cubic centimeter object submerged in mercury? (Your answer should be in newtons.) _____

8. In the figure on the next page, the piston at left has an area of 6 square inches and the one at right 21 square inches. What weight is needed on the left to balance 500 pounds on the right? _____

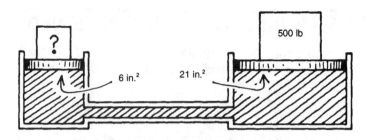

9. What relationship does a molecule of a gas have to its neighboring molecules?

_____

10. Suppose that two gases are in identical containers and that the gases have the same number of molecules with the same average speed. The molecules of one gas are more massive than those of the other, however. Which will exert the greater pressure on the walls of its container? Explain. _____

_____

_____

11. Match the units (on the right) with the quantities that they are used to express. More than one unit may be possible with a given quantity.

   (a) pressure _____

   (b) force _____

   (c) weight density _____

   (d) mass density _____

   lb/in.²   mm Hg

   N/m²   newtons

   kg/m³   lb/ft³

12. When a weather report gives the pressure as 28.9 inches, what is the pressure in lb/in.²? (The density of mercury is 0.49 lb/in.³) _____

13. Will the cornerstone of question 6 apparently weigh more in a vacuum or in air? _____

ANSWERS

1. In a solid the molecules are restricted to definite positions, and they cannot change places with other molecules. In a liquid the molecules exert forces on one another so that distances between molecules are pretty well fixed, but they are able to move around one another. (frame 1)

2. Pressure = depth·weight density (frame 2)

3. 1250 lb/ft² (20 ft·62.4 lb/ft³) (frame 2)

4. 23,520 N/m². Solution: mass density of alcohol = 0.80·1000 kg/m³ = 800 kg/m³
   weight density of alcohol = 800 kg/m³·9.8 N/kg = 7840 N/m³
   pressure = depth·weight density
   = 3 m·7840 N/m³ = 23,520 N/m² (frame 4)

5. 32.4 lb. Solution: The volume of water displaced is 0.52 ft^3. The weight of this much water is 0.52 ft$^3 \cdot$62.4 lb/ft^3 = 32.4 lb. (frames 5, 6)

6. 300 lb. Solution: volume of block = 2 ft^3

   weight of water displaced = 2 ft$^3 \cdot$62.4 lb/ft^3 = 125 lb

   apparent weight = 425 – 125 = 300 lb (frame 5)

7. 13.3 newtons. Solution: The density of mercury is 13.6 times the density of water. This is 13.6 g/cm^3. The mass of mercury displaced by the 100-cm^3 object is 1360 grams, or 1.36 kg. The weight of 1.36 kg is (1.36 kg$\cdot$9.8 N/kg), or 13.3 N. (frame 8)

8. 143 lb. Solution: The pressure on the right side is (500 lb)/(21 in.^{2}) = 23.8 lb/in.^{2}. This pressure is exerted on the left piston, so the force exerted there is

   force = 23.8 lb/in.$^2 \cdot$6 in.^{2} = 143 lb (frame 9)

9. Hardly any! Molecules of a gas exert forces on one another only when collisions occur. (frame 10)

10. the gas with larger molecules; Pressure is due to collisions with the walls. If the molecules are moving at the same speed, the more massive molecules will exert more force in a collision. (frame 12)

11. pressure: lb/in.^{2}, N/m^2, and mm Hg (frames 2, 4, 16)

    force: newtons (frame 4)

    weight density: lb/ft^3 (frame 4)

    mass density: kg/m^3 (frame 4)

12. 14.2 lb/in.^{2}. Solution: pressure = depth$\cdot$weight density

    = 28.9 in.$\cdot$0.49 lb/in.^{3}

    = 14.2 lb/in.^{2} (frame 17)

13. in a vacuum (When in air, it is buoyed up slightly—0.15 lb—by the buoyant force of the air.) (frames 18, 19)

8 Temperature and Heat

Prerequisites: Chapters 5, 6 (frame 1), and 7 (frames 1 and 10)
Before we can define temperature or discuss how it is measured, we need to talk about three more-or-less common temperature scales: Fahrenheit, Celsius (or Centigrade), and Kelvin (or Absolute). Then after seeing how temperature is measured, we will be ready to discuss the real meaning of the term and distinguish between temperature and heat (which are often confused in everyday language).

OBJECTIVES

After completing this chapter, you will be able to

- specify reference points for the three most common temperature scales;
- convert readings from any of the three temperature scales to any other;
- explain the difference in operation between a mercury thermometer and one made with a bimetallic strip;
- using a table of thermal coefficients, calculate change in length due to a given change in temperature of a given solid;
- differentiate between temperature, heat, and internal energy;
- calculate the number of calories or Btu's needed to raise water a given number of degrees;
- use a table of specific heat to calculate how much heat is needed to cause a given temperature change in a given substance.

1 THE FAHRENHEIT TEMPERATURE SCALE

The Fahrenheit temperature scale is defined on the basis of two points: the freezing point of pure water is set at 32°F, and the boiling point of water is 212°F.

(a) How many Fahrenheit degrees* are there between the freezing and boiling points of pure water? _____

(b) What is the status of water at 0°F? _____

Answers: (a) 180;   (b) it is ice (32 degrees below freezing)

---

2 THE CELSIUS SCALE

The Celsius (or Centigrade) scale, although just as arbitrarily defined as the Fahrenheit scale, is part of the metric system. It is the only worldwide system. It sets its two fixed points at 0°C for the freezing point of water and 100°C for the boiling point.

(a) How many Celsius degrees are there between the freezing point and the boiling point of water? _____

(b) Compare your answer to part (a) here to that in frame 1. Which is larger—a Fahrenheit degree or a Celsius degree? _____

Answers: (a) 100;   (b) a Celsius degree

3

A Celsius degree is equal to $\frac{180}{100}$, or $\frac{9}{5}$, times a Fahrenheit degree. Knowing this, the only other thing that one must consider in order to convert from one scale to the other is that there are 32 Fahrenheit degrees difference in the location of the freezing point. To convert °F to °C, one may use the following equation:

$$°C = (°F - 32) \cdot \frac{5}{9}$$

Rewriting this to convert °C to °F:

$$°F = \left(°C \cdot \frac{9}{5}\right) + 32$$

(a) What is 77°F on the Celsius scale? _____

(b) What is 35°C on the Fahrenheit scale? _____

(c) What is −40°C on the Fahrenheit scale? _____

---

*"Degrees Fahrenheit" refers to the actual reading on the temperature scale. "Fahrenheit degrees" refers to a number of degrees of change. Thus, there are 5 Fahrenheit degrees between 40 degrees F and 45 degrees F.

Answers: (a) 25°C. Solution: $(77 - 32) \cdot \frac{5}{9} = 45 \cdot \frac{5}{9} = 25$; (b) 95°F. Solution: $(35 \cdot \frac{9}{5}) + 32 = 63 + 32 = 95$ (c) −40°F. Solution: $(-40 \cdot \frac{9}{5}) + 32 = -72 + 32 = -40$ (This is the only place where °C = °F!)

4 | THE KELVIN SCALE

It can be shown that the coldest possible temperature is −273.15°C. This leads to our last scale, which has its zero point located at this coldest temperature. Then, rather than define another fixed point, we simply say that one division on this scale is equal to one division on the Celsius scale. Thus, the freezing point of pure water on the Kelvin scale is 273.15 K. (There is no degree symbol used with this scale; 290 K is read "290 Kelvins.")

The conversion between Celsius and Kelvin is easy: One simply adds 273.15 to the Celsius temperature.

(a) What is the boiling point of water on the Kelvin scale? _____

(b) What is 95°F on the Kelvin scale? _____

(c) What is 500 K on the Celsius scale? _____

Answers: (a) 373.15 K; (b) 308.15 K (First, change to °C: 35°C.); (c) 226.85°C

5

The Kelvin scale is not quite so arbitrary as the other two scales, because it starts at a real starting point in nature, the coldest possible temperature. The Kelvin scale is seldom used, however, outside the scientific community. The figure at right compares the three scales.

Use the figure to determine the following.

(a) What is 32°F on the Celsius scale? _____
On the Kelvin scale? _____

(b) What temperature is 22°C on the Fahrenheit scale? _____

(c) What is the Fahrenheit reading of the coldest temperature possible? _____

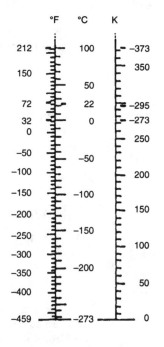

Answers: (a) 0°C, 273 K; (b) 72°F—a comfortable temperature; (c) −459°F

6 | EXPANSION RELATED TO TEMPERATURE: THERMOMETERS

Most materials expand when their temperature increases. Allowance for this expansion is made along the sides of long brick buildings by locating spaces every so often for the brick to expand into during hot weather. (Such spaces are usually filled with a relatively soft caulk.) What is not quite so well known is that different materials expand different amounts with the same temperature rise. This fact allows the common mercury thermometer to register temperature.

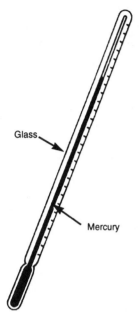

In the figure, consider what happens when the temperature of the materials of the thermometer rises. Both the glass and the mercury expand, but the mercury expands much more. If the mercury in the bulb is to expand, it must move up the thin column. Marks along the column indicate the temperature.

(a) If glass expanded more than mercury when they are heated, what would happen to the level of the mercury in the column?

_____

(b) When the temperature falls, mercury contracts, again more than the glass does. What happens to the mercury in the column now?

_____

Answers: (a) It would descend.   (b) It falls.

7

The upper figure in this frame shows two different metals connected together to form a **bimetallic strip.** Suppose the upper metal expands more than the bottom one when they are heated. The strip will bend as shown in the lower figure. Such bimetallic strips are at the heart of oven thermometers and thermometers found in most wall thermostats.

(a) What would happen if the two metals expanded the same amount when heated? _____

(b) How would the strip bend if the lower metal expanded more than the top one when they are heated? _____

_____

Answers: (a) The strip wouldn't bend.   (b) It would curve upward—with the lower strip on the outside of the curve.

8

The amount of expansion of most materials is very nearly proportional to the rise in temperature. For example, if an iron rod expands 1 millimeter when its temperature rises 15°, then it will expand 2 mm with a 30° rise in temperature. The equation used to calculate expansion due to temperature change is:

$$\Delta l = \alpha \cdot l \cdot \Delta T$$

where Δl is the change in length and ΔT is the change in temperature.* The other symbol, α (alpha), is called the thermal coefficient of linear expansion and is a factor that is a property of the material being considered. Table 8.1 gives the coefficients of expansion for a number of common materials.

Table 8.1

Material	α (per C°)
aluminum	0.000025
brass	0.000018
glass (pyrex)	0.000003
iron	0.000011

The other quantity on the right side of the equation above, l, is the original length of the object, and must be taken into consideration.

Suppose we have an iron railroad rail 15 meters long. We will calculate its expansion when the temperature changes from –5°C to 25°C. (This corresponds to a change from 23°F to about 78°F—"everyday" temperatures.)

(a) How much does the temperature change on the Celsius scale? _____

(b) How much does the rail expand? _____

(c) Express this expansion in millimeters. _____

Answers: (a) 30 Celsius degrees; (b) 0.00495 m. Solution: Δl = (0.000011/C°)·15 m·30 C°
(c) 4.95 mm

9

Now you have a brass statue that is 3 feet high. How much will its height increase if, after being at room temperature (about 20°C), boiling water is constantly poured over it? We can use British units for length in this equation because the coefficient of expansion times the temperature change is equal to the ratio of the change in length to the original length. And a ratio has no units.

---

*The symbol Δ, the Greek letter delta, is used to represent a change in a quantity. Thus, on the left side of the equation, Δl means "change in length."

Fill in the values in the equation:

$$\Delta l = \alpha \cdot l \cdot \Delta T$$

$\Delta l =$ _____ $\cdot$ _____ $\cdot$ _____ $=$ _____

Answer: $\Delta l = 0.000018/C° \cdot 3$ ft $\cdot 80$ C° $= 0.00432$ ft (about 0.05 inch)

10 TEMPERATURE, INTERNAL ENERGY, AND HEAT

Thus far we have seen how to measure temperature and how a change in temperature results in a change in the size of objects. What are we actually measuring when we measure temperature? Recall that in all materials, molecules are in constant motion. In a solid, this motion consists of vibration of the molecules, while in fluids the motion consists of the molecules bouncing around from one place to another. Motion, however, implies energy—kinetic energy. Temperature is simply the measure of the average kinetic energy of the molecules. In an object with a greater temperature, the molecules have more kinetic energy and therefore greater speed.

Internal energy refers to the total kinetic and potential energies of the molecules of an object. (The potential energy results from the forces between molecules of the object.)

(a) Which molecules have greater average kinetic energy, those in a cup of water at 20°C or in one at 20°F? _____

(b) Which molecules have greater average kinetic energy, those in a cup of water at 40°C or those in a quart of water at 40°C? _____

(c) What causes the potential energy of the molecules of a substance? _____

_____

Answers: (a) 20°C (because it is the higher temperature); (b) the same; (c) forces between molecules

11

Temperature involves only kinetic energy while internal energy involves both kinetic and potential energy. A second important distinction between the two is that temperature measures the *average* kinetic energy of the molecules, and internal energy refers to the *total* energies of all the molecules of an object. Suppose two cups of water, each at 40°C, are poured together.

(a) What is the temperature (due to the average kinetic energy) of the mixed water? _____

(b) How does the internal energy of the mixture compare to the internal energy of one of the cups before mixing? _____

Answers: (a) 40°C (the same as the temperature of the individual cups); (b) It is twice as much (because it is the total energy).

12   Heat is a term that is often confused with temperature (and sometimes with internal energy). Heat is the internal energy that is transferred from one object to another as a result of a difference in temperature between the two objects. Note that heat is not "in" an object, but is energy that is being transferred from one object to another. Methods of heat transfer are a subject for the next chapter. For now, it is sufficient that you realize that when objects of different temperature are put in contact (or mixed) with one another they come to the same temperature by a transfer of heat from the hot to the cold object.

Identify each description below as representing heat, temperature, or internal energy:

(a) Average kinetic energy of molecules: _____

(b) Transfer of energy: _____

(c) Total energy in an object: _____

Answers: (a) temperature; (b) heat; (c) internal energy

13   THE CALORIE

Although heat and temperature are different things, they are obviously related. Usually, when we add heat to an object, its temperature rises. This leads to the definition of the calorie.* One calorie is the amount of heat which, when added to 1 gram of water, raises the temperature of the water 1 Celsius degree.

(a) Suppose you have 15 grams of water and add 60 calories to it. How much does its temperature rise? _____

(b) How much heat must be added to raise 100 grams of water from 5°C to 95°C? _____

Answers: (a) 4 Celsius degrees (60/15); (b) 9000 calories (The change in temperature is 90°, so it takes 90 calories to change the temperature of each gram of water this much.)

---

*Since heat is energy, the calorie is a unit of energy. When this word is used to rate the energy content of food, what is really meant is the kilocalorie, which is 1000 of the calories we are defining here. It is unfortunate that we refer to the dietary kilocalorie as a calorie, but confusion seldom results since the meaning is usually clear from the context.

14 THE BRITISH THERMAL UNIT

In the British system, the unit of heat is the British thermal unit, the Btu. One Btu is the amount of heat that will raise the temperature of 1 pound of water 1 Fahrenheit degree. Note the parallel in the definitions of the calorie and the Btu.

You may hear of an air conditioner rated as, say, 12,000 Btu's. This makes no sense! Are we to suppose that the air conditioner is capable of exchanging 12,000 Btu's of heat energy before it "expires"? No. In fact, the rating is 12,000 Btu's per hour. The last two words are usually omitted, probably because so few people know what is meant anyway. Ask someone who sells air conditioners.

(a) How many Btu's are required to raise the temperature of 5 pounds of water 10 Fahrenheit degrees? _____

(b) What does a rating of 12,000 Btu's per hour mean on a heater or air conditioner? _____

Answers: (a) 50 Btu's; (b) 12,000 Btu's of energy can be exchanged each hour by the device.

15 SPECIFIC HEAT

When 1 calorie of heat is added to 1 gram of water, the water's temperature rises 1 Celsius degree. To raise 1 gram of glycerin 1 degree, on the other hand, requires only 0.60 calorie. And to raise 1 gram of aluminum 1 degree requires only 0.22 calorie. Specific heat is defined by the following equation:

$$\text{Specific heat} = \frac{\text{the heat needed to raise 1 gram of a substance 1 degree}}{\text{the heat required to raise 1 gram of water 1 degree}}$$

Table 8.2 lists the specific heats of a number of substances.

Table 8.2

Material	Specific heat
aluminum	0.22
copper	0.09
glass	0.16
glycerin	0.60
ice	0.50
iron	0.11
silver	0.06
steam	0.50
water	1.0000

Since water requires 1 calorie per gram Celsius degree, the specific heat of a substance is equal to the number of calories required to raise the temperature of 1 gram of the substance 1 Celsius degree. The relationship can best be expressed as a formula:

$$H = m \cdot c \cdot \Delta T$$

where H is the heat added (or removed), m is the mass of the substance, c is its specific heat, and ΔT is the change in temperature.

(a) How much heat is required to raise the temperature of 150 grams of iron from 20°C to 25°C? _____

(b) Which of the materials of Table 8.2 will show the greatest increase in temperature if the same amount of heat is added to 1 gram of each of them?

_____

(c) If 120 calories of heat are added to 100 grams of glass at 15°C, what will be the final temperature of the glass? _____

Answers: (a) 82.5 cal. Solution: $H = m \cdot c \cdot \Delta T$
$$= 150 \cdot 0.11 \cdot (25 - 20)$$
$$= 82.5 \text{ cal}$$

(b) silver (Silver requires the least amount of heat to raise its temperature a given amount. Thus, if the same amount of heat is added to each material, the silver will show the greatest temperature change.)

(c) 22.5°C. Solution: $H = m \cdot c \cdot \Delta T$
$$\Delta T = \frac{H}{m \cdot c}$$
$$= \frac{120 \text{ cal}}{100 \text{ g} \cdot 0.16}$$
$$= 7.5 \text{ degrees}$$

Thus, the new temperature = 15° + 7.5° = 22.5°

SELF-TEST

1. State the freezing point and the boiling point of water on each of the three common temperature scales. _____

_____

_____

2. Change 68°F to its Celsius temperature. _____

3. Forty degrees Celsius is how many degrees Fahrenheit? _____

4. Forty degrees Celsius is how many Kelvins? _____

5. Suppose a liquid-in-glass thermometer were constructed with a liquid that has the same coefficient of expansion as the glass. What would happen when the

temperature of the thermometer is increased? _____

6. How does a bimetallic strip respond to temperature change? _____

7. How much will a 200-ft aluminum wire expand if it is heated from 20°C to 60°C? (Use Table 8.1.) _____

8. Will the aluminum wire of question 7 expand more or less than a brass wire of the same length with the same temperature change? (Use Table 8.1.)

   _____

9. If 30 Btu's of heat are added to 5 pounds of water, how much will the temperature of the water change? _____

10. How much heat must be added to 100 grams of copper to raise its temperature 5 Celsius degrees? (Use Table 8.2.) _____

11. How does the amount of heat required to raise the temperature of water a certain amount compare to the heat required to raise the same mass of ice the same amount? (Use Table 8.2.)_____

   _____

   _____

ANSWERS

1. | Scale | Freezing Point | Boiling Point |
   |---|---|---|
   | Fahrenheit | 32° | 212° |
   | Celsius | 0° | 100° |
   | Kelvin | 273.15° | 373.15° |

   (frames 1, 2, 4, and 5)

2. 20°C. Solution: $(68 - 32) \cdot \frac{5}{9} = 20$ (frame 3)

3. 104°F. Solution: $(40 \cdot \frac{9}{5}) + 32 = 104$ (frame 3)

4. 313 K. Solution: $40 + 273.15 = 313.15$ (frame 4)

5. The liquid would neither rise nor fall, because it would expand just as much as the glass. (frame 6)

6. The strip bends so that the metal that expands more is on the outside of the curve. (frame 7)

7. 0.2 ft. Solution: $\Delta l = \alpha \cdot l \cdot \Delta T$
   $$= 0.000025 \cdot 200 \cdot (60 - 20) \text{ (frame 8)}$$

8. more (because the aluminum's coefficient of expansion is greater) (frame 8)

9. 6 Fahrenheit degrees (Five pounds of water requires 5 Btu's to raise it one degree; 30 Btu's raise it six times as far—6 degrees.) (frame 14)

10. 45 cal. Solution: $H = m \cdot c \cdot \Delta T$
    $$= 100 \cdot 0.09 \cdot 5 \text{ (frame 15)}$$

11. It requires twice as much heat to raise water's temperature as it does to raise ice's (assuming the same amount of each substance is to be raised by the same temperature), because the specific heat of water (1.00) is twice that of ice (0.50). (frame 15)

9 Change of State and Transfer of Heat

Prerequisites: Chapters 5–8

In the last chapter we saw that, given the specific heat of a substance, we can calculate how much heat is required to raise the substance's temperature a certain amount. We ignored the fact that, if enough heat is added, a solid may melt or a liquid may vaporize. We will take up these phenomena in this chapter, and then will discuss how heat is transferred from one object to another.

OBJECTIVES

After completing this chapter, you will be able to

- specify the heat of fusion of water;
- specify the heat of vaporization of water;
- calculate the amount of heat required to change ice, water, or vapor at any temperature to any other state or temperature;
- explain why amorphous solids have no heat of fusion;
- differentiate among the three ways that heat is transferred;
- calculate the heat conducted through a given object, given its conductivity, its dimensions, the time elapsed, and the temperature differential;
- relate the color of an object to its ability to absorb and emit thermal radiation;
- identify the predominant method of heat transfer involved in various situations.

1 MELTING: THE HEAT OF FUSION

The freezing point of water is 0°C. At this temperature, water may be a liquid or it may be a solid—ice. The amount of internal energy determines in which of these states the water will be. If we add heat to ice that is at the freezing point (or melting point, which is the same thing), the ice will become water at that same temperature. The amount of heat required to change 1 gram of ice at 0°C to

1 gram of water at 0°C is 80 calories. This is called the **heat of fusion** of water. Other substances generally have a lower heat of fusion than water does. For example, lead's heat of fusion is 5.9 calories per gram.

(a) How many calories are needed to change 10 grams of ice at 0°C to 10 grams of water at 0°C? _____

(b) How many calories are needed to change 10 grams of solid lead at its melting point to liquid at that temperature? _____

Answers: (a) 800;   (b) 59 (Thus, lead is easier to melt—once you reach its melting point of 327.5°C or 621°F.)

2

In a crystalline solid (the type we have been discussing), the atoms are held in a definite pattern by forces exerted by neighboring molecules. Since the same pattern of molecular arrangement is repeated throughout the solid, the bonds holding one molecule in place are similar to the bonds holding others. This results in the same energy being required to break one bond as to break any other. As a solid is heated, the molecules vibrate with more and more energy (recall that temperature is the average kinetic energy of the molecules), and at some point any additional energy will begin to break their bonds. The energy that is required to break these bonds is the heat of fusion.

An amorphous solid (such as butter or glass) does not have a regular pattern of molecular arrangement, so different molecules are held with forces of different strength. Thus, as the solid is heated, some bonds break at low temperatures and some at higher temperatures. This results in the gradual softening of such substances rather than a sudden melting at a given, fixed temperature (as is the case with crystalline solids).

(a) Based on your experience, do you expect margarine to be a crystalline or amorphous solid? _____

(b) Which have no heat of fusion, crystalline or amorphous solids? _____

(c) Which have molecular bonds of varying strengths, crystalline or amorphous solids? _____

Answers: (a) amorphous (since it softens gradually rather than melting at a definite temperature);   (b) amorphous;   (c) amorphous

3   THE HEAT OF VAPORIZATION

Suppose you start with water at room temperature—about 20°C—and you add heat. Each calorie of heat added will increase the temperature of a gram of water 1 Celsius degree. Then when the water reaches 100°C, it will not increase further in temperature, but will start to boil. Boiling is the process by which the water changes to its gaseous state, water vapor. To change 1 gram of water at 100°C to water vapor at 100°C requires 540 calories of heat.

(a) Which requires more calories per gram, melting ice, or vaporizing water?

_____

(b) How much heat is needed to change 10 grams of water at 100°C entirely to water vapor? _____

Answers: (a) vaporizing water;   (b) 5400 calories

4 Let's try a more complex problem: Given that the specific heat (*c*) of ice is 0.5, and the specific heat of water vapor is also 0.5, how much heat is required to change 1 gram of ice at –20°C to water vapor at 130°C? Fill in the steps below.

(a) The first step is to find out how much heat is needed to change the ice at –20°C to ice at 0°C. Use the equation $H = m \cdot c \cdot \Delta T$. What is your answer?   _____ calories

Total so far: (_____)

(b) For the next step, change the ice to water at the same temperature.   _____ calories

Total so far: (_____)

(c) Now heat the gram of water from 0°C to 100°C.   _____ calories

Total so far: (_____)

(d) The next step is to vaporize the water.   _____ calories

Total so far: (_____)

(e) Finally, use the equation in (a) to calculate the heat needed to raise the temperature of the steam from 100°C to 130°C.   _____ calories

(f) What is the total number of calories needed?   _____ calories

Answers: (a) 10 (10) (Chapter 8, frame 15);   (b) 80 (90);   (c) 100 (190);   (d) 540 (730);
(e) 15;   (f) 745

5 In a liquid, molecules are free to move around one another, but there are forces that prohibit them from separating from one another. At any given temperature, the molecules of a liquid have a certain average kinetic energy. I emphasize that this is an *average* energy, however. As they bounce around each other, some molecules momentarily have a greater-than-average energy and some have less-than-average energy. Sometimes a molecule with a lot of energy hits the surface of the liquid and breaks free from the surface. This is the process of evaporation.

(a) Which would you expect to evaporate more quickly, a warm liquid or a cool one? _____

(b) What difference in the molecules causes this difference in evaporation rate?

_____

Answers: (a) warm liquid;   (b) There are more fast-moving molecules in the warm liquid.

6

Now suppose we get the liquid hot enough that molecules have enough energy to break free from one another down within the liquid rather than just on the surface. This is boiling. When a liquid boils, the molecules within the liquid break from one another, forming bubbles of vapor, which rise to the surface. To break the molecules free from one another requires much more energy than is required to break them from fixed positions in a solid, and thus the heat of vaporization of water is much greater than the heat of fusion (540 calories per gram rather than 80).

(a) What is the name of the amount of heat required to break all molecules from a fixed position in a gram of a substance? _____

(b) What is the name of the amount of heat required to break all molecules free from one another in a gram of a substance? _____

(c) What is happening at the molecular level as a liquid boils? _____

_____

Answers: (a) heat of fusion;   (b) heat of vaporization;   (c) Molecules are breaking free from one another below the surface of the liquid.

---

7 TRANSFER OF HEAT: CONDUCTION

If you stick the end of a poker into a fire and hold the other end, you will soon feel your end getting hotter. Heat is being transferred from the hot end to the cold end by the method of **conduction**. Picture the molecules in the hot end of the poker in rapid vibration while the molecules of the cold end are vibrating less rapidly. Since there are forces between the molecules, a rapidly vibrating molecule will transfer some of its vibrational energy to its slower moving neighbor. In this way, the kinetic energy of vibration of the molecules is gradually spread throughout the metal until the molecules at one end have about the same average kinetic energy as the molecules at the other.

Because of the differences between bonding forces in various materials, some materials are better conductors of heat than others. In general, metals are good conductors of heat, and metals that conduct electricity best also conduct heat best. Copper, for example, is a better conductor of both electricity and heat than is aluminum, and silver is better than either. Some materials conduct heat so poorly that we classify them as heat insulators. Wood is only 0.1% as effective in conducting heat as is aluminum, and is a fair insulator.

In your experience, which conducts heat more rapidly, iron or wood?_____

Answer: iron

8

Suppose we wish to calculate the amount of heat that will flow through a window pane on a cold winter day. The variables that must be considered are listed below.

1. The thickness of the glass. If the glass is thicker, less heat will flow. (We'll call this *d*, for distance.)

2. The area of the pane of glass. The greater the area, the more heat flows. (*A* = area.)

3. The difference in temperature between the two surfaces. A greater difference will cause a greater heat flow. (ΔT = temperature difference.)

4. The time, *t*, during which the heat flows.

5. The thermal conductivity, *k*. The accompanying table shows heat conductivities of various materials. The units are cal/cm·s·C°.

The equation relating all of these variables is as follows:

$$H = \frac{k \cdot t \cdot A \cdot \Delta T}{d}$$

The units of *H*, of course, are calories.

Table 9.1

Substance	Thermal conductivity (cal/cm·s·C°)
silver	0.97
copper	0.92
aluminum	0.50
iron	0.16
glass	0.0025
wood	0.0005
felt	0.00004

Now suppose the pane is 15 cm by 20 cm and is 0.3 cm thick. The temperature of one side of the glass is 5 Celsius degrees warmer than the other side. How much heat escapes through the window in 1 hour? _____

Answer: 45,000 calories

Solution: $H = \dfrac{0.0025 \ (\text{cal/cm·s·C}°)·3600 \ \text{s}·300 \ \text{cm}^2 \ 5·\text{C}°}{0.3 \ \text{cm}}$

9 CONVECTION

Convection, the primary method by which heat moves through a fluid, is the movement of heat by transfer of the molecules of the fluid itself. As a substance becomes warmer, it expands. When a solid does this, the molecules have to stay in their relative positions, but the expansion of part of a liquid or a gas causes that part of the substance to be less dense than its surroundings. When part of the fluid becomes less dense than the surrounding fluid, it rises. (We sometimes say that "heat rises," but what we mean is that hot fluids rise through cooler fluids of the same substance.) You can see convection currents if you place a pan of water on a hot stove and put a drop or two of food coloring in the water. The stove burner causes the water on the bottom of the pan to increase in temperature. The newly heated water is less dense than the cooler water above, so it rises. The food coloring allows you to see the currents within the water as the heat is spread throughout.

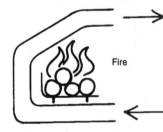

Fire

(a) What do convection currents do to the process of diffusion?

_____

(b) Pipes such as those shown in the figure are sometimes put in fireplaces to increase the heat output of the fireplace. What method do these pipes use to get heat from the fireplace into the room? _____

Answers: (a) speed it up; (b) convection (Heat is transferred from the fire through the walls of the pipes by conduction. Convection then brings the heat out into the room.)

10 RADIATION

Heat can also be transferred from one place to another by radiation. Every object is constantly absorbing and emitting radiation, but we do not notice it until the object is hot enough to emit thermal radiation, to which our skin is sensitive. Such thermal, or infrared, radiation is electromagnetic in nature and travels at the speed of light. Radiant heat is easily felt in front of a fire, a heat lamp, or even near a regular lightbulb.

An object that is a good absorber of radiation is also a good radiator. The best absorber (and therefore the best radiator) is a black object. An object with a shiny, silvery surface neither absorbs nor emits radiation very effectively. This is why new houses are wrapped in a shiny foil as part of their insulation.

(a) Why are white clothes recommended for hot, sunny locations? _____

_____

(b) By what method is heat transferred from sun to earth? _____

Answers: (a) White clothes do not absorb thermal radiation from the sun as much as dark-colored clothes. (b) radiation (The other two methods can play no part out in space because there are no atoms present to conduct or convect heat.)

11

A vacuum flask (popular brand name: Thermos®) is constructed so as to reduce heat loss by conduction, convection, and radiation. The figure illustrates the glass within a vacuum flask.

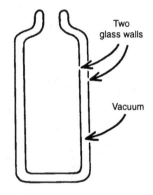

Between the two walls is a vacuum. Since a vacuum contains no material, it can neither conduct nor convect heat. Radiation is reduced by silvering the walls of the glass.

(a) Many old homes are heated by hot water (or steam) "radiators." These work on the same principle as the "radiator" in a car. By what method of heat transfer is most heat carried from such devices? (Hint: Do not be misled by the names.) _____

(b) Why do astronauts wear shiny suits during spacewalks? _____

_____

(c) When you touch a metal lamp post in cold weather, by what method is heat transferred from your hand? _____

Answers: (a) convection (Air moves across the hot surface of the "radiator," carrying away heat. If radiation were the primary method of heat transfer, the devices should be painted black!); (b) to reflect infrared radiation from the sun; (c) conduction

SELF-TEST

Refer to the table in frame 8 at any time.

1. How many calories of heat must be removed from 5 grams of water at 0° Celsius to change it into ice at 0°C? _____

2. How much heat is needed to change 5 grams of water at 90°C to steam at 100°C? _____

3. Explain, on the molecular level, what the energy is used for when heat is added to ice to change it into water. _____
_____
_____

4. Explain why an amorphous solid does not have a definite melting point and a heat of fusion. _____
_____
_____

5. Why is the heat of vaporization of water so much greater than the heat of fusion? _____
_____
_____

6. What are the three methods by which heat is transferred? _____
_____

7. Which is the best conductor of heat—iron, silver, or copper? (Use the table of frame 8.) _____

8. An iron rod has a cross-sectional area of 1 square centimeter and is 1 meter long. It is placed so that one end is in a fire at 700°C and the other end is kept in water at 50°C. How much heat flows through the rod in 1 minute?
_____

9. Gold melts at 1063°C. Is it an amorphous or crystalline solid? _____

10. White lines are often painted across asphalt streets in beach areas. Why are the lines cooler to walk on than the asphalt? _____
_____

11. If a wooden bench and an iron post are both at –5°C (on a cold morning, perhaps), why does the iron post feel cooler to the touch? _____

_____

ANSWERS

1. 400 calories (5 grams·80 calories per gram) (frame 1)

2. 2750 calories. Solution: To change the temperature from 90°C to 100°C requires 50 calories (calculated from $H = m \cdot c \cdot \Delta T$), and to vaporize 5 grams of water requires 5·540, or 2700 calories. Total: 2750 calories. (frames 2, 3)

3. The energy is used to break the bonds that hold the molecules of the solid ice in their fixed positions. (frame 2)

4. In an amorphous solid, the individual molecules have bonds of different strengths holding them in place. Thus, different energies are required to break the bonds and they therefore break at different temperatures. The solid softens as it is heated rather than melting at a definite temperature. Since it does not melt, it has no heat of fusion. (frame 2)

5. In vaporization, the molecules must completely break free of one another rather than just loosen their bonds so that they can move around one another. (frame 6)

6. conduction, convection, and radiation (frames 7, 9, 10)

7. silver (The table shows its thermal conductivity to be greatest.) (frame 8)

8. 62.4 calories. Solution:
$$H = \frac{k \cdot t \cdot A \cdot \Delta T}{d}$$
$$= \frac{0.16 \cdot 60 \cdot 1 \cdot 650}{100}$$
$$= 62.4 \quad \text{(frame 8)}$$

9. crystalline (because it has a definite melting point) (frame 2)

10. They do not absorb as much thermal radiation from the sun, so they are cooler. (frame 10)

11. The iron conducts heat away from our hand much faster than does the wood because iron is a better heat conductor than wood. (frames 7, 8)

<u>10</u> Wave Motion

No prerequisites

Wave motion is basic to an understanding of various aspects of physics, including sound, light, and parts of electricity and magnetism. This chapter will explore wave motion, and the next two will apply the concepts of this chapter to the subject of sound.

OBJECTIVES

After completing this chapter, you will be able to

- determine the relative periods of simple pendula of different lengths and masses;
- specify the effect of a change of length, mass, or amplitude on the period of a pendulum;
- specify the relationship between frequency and period, and—given one—calculate the other;
- describe a wave in terms of wavelength, amplitude, frequency, and speed;
- use the equation $v = \lambda f$ to determine the velocity of a wave;
- differentiate between transverse and longitudinal waves;
- identify energy as being transmitted by a wave;
- describe the effect of relative motion between the sender and the receiver of a wave;
- describe a bow wave and state the conditions necessary to achieve one;
- given one of the two, determine the length or period of a simple pendulum (optional).

1 THE PERIOD OF A PENDULUM

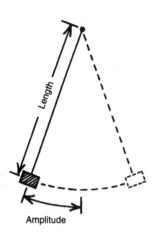

Length

Amplitude

There are various types of pendula (the plural of pendulum). A board swinging from one end is a pendulum, and so is a tire swinging on a rope. We will discuss only what is called a "simple pendulum." A simple pendulum consists simply of a suspended object (the pendulum "bob") that swings freely on a string. The bob must be very small compared to the length of the string and the string must be able to be considered massless in comparison to the mass of the bob. The **period** of a pendulum is the time taken to complete one back-and-forth swing. For example, the period of the pendulum that swings beneath a large grandfather clock is usually 2 seconds. To a high degree of precision, the period depends only upon the length of the pendulum and not upon the angle of swing. (This is true as long as the angle of swing does not exceed 5 or 6 degrees. As the angle—the amplitude—gets larger, the period does also, but the effect is hardly measurable unless large angles are used.) The longer the length of a pendulum, the longer its period.

(a) If two pendula have the same length, but one has a 2-pound weight on its end while the other has only a 1-pound weight, which will have the longer period? _____

(b) A pendulum with a suspending string 2 meters long has a mass of 10 kilograms. Another with a string 4 meters long has a mass of 5 kg. Which will have the longer period? _____

(c) If the period of a pendulum is 1 second, how long does it take to go from one end of its swing to the other? _____

Answers: (a) the same;   (b) 4 meters long;   (c) $\frac{1}{2}$ second

2 RELATION BETWEEN LENGTH AND PERIOD (OPTIONAL)

The period of a simple pendulum is given by the following equation:

$$t = 2\pi \sqrt{\frac{l}{\mathbf{g}}}$$

where t is the period, l is the length of the pendulum (measured to the center of mass of the pendulum bob), and $\mathbf{g}$ is the acceleration of gravity (about 9.8 m/s² or 32 ft/s² at the earth's surface—see Chapter 1, frames 3–7).

(a) By what factor must you increase the length in order to double the period of a pendulum? _____

(b) Use the equation to calculate the length needed to obtain a period of 1 second. (Hint: Start by squaring both sides of the equation.) _____

Answers:
(a) 4 times (since the period depends upon the square root of the length);
(b) 0.25 meters or 0.81 feet

Solution: $t^2 = 4\pi^2 \cdot \dfrac{l}{g}$

$$l = \dfrac{t^2 g}{4\pi^2}$$
$$= \dfrac{(1 \text{ s})^2 \cdot 9.8 \text{ m/s}^2}{4 \cdot 3.14^2}$$
$$= 0.25 \text{ m}$$

(Using 32 ft/s² for the acceleration of gravity yields 0.81 feet as the length.)

3   FREQUENCY

The **frequency** of a pendulum (or of any regularly repeating motion) is the number of cycles (periods) completed in a unit of time. The 2-second period of the grandfather clock pendulum corresponds to a frequency of $\frac{1}{2}$ cycle per second or 30 cycles per minute. The simple relationship between period and frequency is shown below:

$$f = \frac{1}{T}$$

where *f* is frequency and *T* is period (abbreviated as it is because period is a time)*.

(a) What is the frequency of a pendulum that has a 1-second period?

_____

(b) What is the frequency of a pendulum having a period of $\frac{1}{3}$ second?

_____

(c) What is the period of a pendulum having a frequency of 2 cycles/second?

_____

(d) Which has the greater frequency, a pendulum of length 1 foot or one of length 3 feet? _____

Answers: (a) 1 cycle/second (or 60 cycles/minute);   (b) 3 cycles/second;   (c) $\frac{1}{2}$ second;
(d) 1 foot (because it has the smaller period—see frame 1)

---

*The expression 1/*T* is called the reciprocal of *T*. Frequency is the reciprocal of period.

4 WAVES

Suppose a pendulum is arranged so that at one end of its swing it hits the end of a long tank of water. And suppose that somehow the pendulum does not decrease its amplitude, so it continues to hit the tank with each swing. Each time the tank is hit, a small ripple will be sent down the surface of the water.

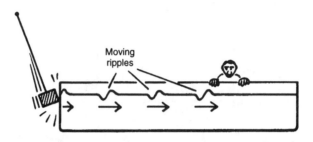

Now if the pendulum is from the grandfather clock, with its 2-second period, these ripples pass a person standing by the tank with the same frequency that the pendulum has.

(a) What is the frequency of the wave caused by the grandfather clock pendulum?

_____

(b) If the period of a wave is the time between successive high points of the wave as they pass a given location, what is the period of this wave? _____

Answers: (a) $\frac{1}{2}$ wave/second*;   (b) 2 seconds

5

Three more variables enter the picture in describing period and frequency of waves. The velocity or speed of a wave is defined in exactly the same way as is the speed of a material object: the distance traveled divided by the time of travel. The **wavelength** is the distance between two consecutive corresponding points on the wave. In the figure of this frame, the wavelength of the wave is the distance from X to Y, or from A to C.

A final variable is **amplitude**. Amplitude of a wave is defined as the distance from the center, rest position of the wave to the point of maximum displacement. In the figure the amplitude is the distance M. (Note that the amplitude is *not* distance N. N is twice the amplitude.)

---

*Notice that the unit has changed from "cycle/second" to "wave/second." Often the unit of frequency is simply stated as "/second" or "per second." Another name for this unit is "hertz," so that 0.5 wave/second is 0.5 hertz.

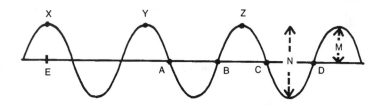

(a) Which of the variables described in this frame are labeled with letters in the figure? _____

(b) What is the distance from B to D called? _____

(c) How is the speed of a wave defined? _____

_____

(d) What fraction of a wavelength is the distance from A to B? _____

(e) What is the distance from E to X called? _____

Answers: (a) wavelength and amplitude;   (b) one wavelength;   (c) the distance traveled divided by the time of travel;   (d) $\frac{1}{2}$;   (e) amplitude

---

6   RELATIONSHIP BETWEEN THE VARIABLES

To see the relationship between velocity, frequency, and wavelength, imagine yourself on a fishing dock observing waves passing under you. You observe that 30 waves pass you per minute, and you discover (with a handy meter stick) that each wave is 2 meters long. Now what is the speed of the waves? Since 30 waves come by each minute and each wave is 2 meters long, the speed is 60 meters/minute.

In general, we multiply frequency times wavelength to obtain velocity. Letting the Greek letter lambda (λ) represent wavelength, we have:

$$v = \lambda f.$$

(a) What is the wavelength in the example in which you were on the fishing dock? _____

(b) What is the frequency? _____

(c) If the speed of the waves remains the same, but the frequency is doubled, what happens to the wavelength of the waves? _____

Answers: (a) 2 meters;   (b) 30 per minute;   (c) It is halved.

7

This equation that relates velocity, wavelength, and frequency applies to all types of waves, including water waves, sound waves, and light waves. Try a few more applications of $v = \lambda f$.

(a) Sound waves travel at about 1100 ft/s in air. What is the wavelength of a sound that has a frequency of 550 cycles per second? _____

(b) Suppose the period of some small ripples on water is $\frac{1}{10}$ second, and their wavelength is 2 centimeters. What is the velocity of the waves? _____

Answers: (a) 2 ft (which is $\dfrac{1100 \text{ ft/s}}{550/\text{s}}$);   (b) 20 cm/s (Remember, $f = 1/T$.)

---

8   ## TRANSVERSE AND LONGITUDINAL WAVES

Thus far we have spoken of waves that vibrate in a direction perpendicular to the direction the wave is traveling. For example, the primary motion of water on the surface is up and down as a wave passes. (We are speaking of smooth waves away from the shore; waves breaking on a beach are another matter.) If two girls hold opposite ends of a long spring and one vibrates her end up and down, waves similar to the water waves travel down the length of the spring. Again, the primary motion of the coils of the spring is up and down although the wave itself is moving horizontally. Waves in which the motion of the particles is perpendicular to the motion of the wave are called **transverse waves.**

Now suppose that the two people again hold opposite ends of a spring. (If you want to try this, a Slinky® spring will work best here.) This time one of the people squeezes together a number of coils and releases them. Soon after release the spring will look somewhat like that in the figure.

Here the squeezed-up portion (called a **compression**) travels along the spring. The motion of an individual coil as the wave passes is back and forth along the direction of the wave's velocity. Such a wave is called a **longitudinal wave.** If instead of compressing the coils, the person had stretched them apart, a **rarefaction** would have been produced, and would also have moved along the wave as did the compression. (A rarefaction is an area where the coils are stretched out.)

Although transverse and longitudinal waves seem very different in nature, they both obey the basic rules of wave motion, and, of course, the $v = \lambda f$ equation

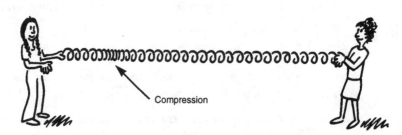

Compression

applies to both. We will see that while light acts as a transverse wave, sound waves are longitudinal.

(a) In what type wave is the vibration perpendicular to the direction of travel of the wave? _____

(b) What type vibrates parallel to the direction of travel? _____

(c) What type wave contains compressions and rarefactions? _____

(d) What type wave is produced when you move one end of a horizontal spring up and down? _____

(e) What type wave has a wavelength? _____

Answers: (a) transverse; (b) longitudinal; (c) longitudinal; (d) transverse; (e) both

9 ENERGY TRANSFER

Consider what is moving as a wave moves along a spring. The wave itself moves from one end of the spring to the other, but the individual coils of the spring simply vibrate about a center position. The vibration may be up and down or back and forth, but in neither case do the coils themselves move down the spring. The same idea holds in the case of all wave motion. Waves do not transmit matter; they transmit energy.

We have defined energy (in Chapter 3, frame 17) as the ability to do work and work as the product of force and distance (Chapter 3, frame 13). Now consider a transverse pulse moving along a spring toward the person holding one end. When the pulse reaches him, he feels a force pulling up (or down) on his hand, and the force will move his hand slightly. Thus, it does work on his hand. The person who started the wave pulse put energy into the spring (by exerting a force on it through a distance as he wiggled it), and this energy traveled in the form of a wave to the other end. Unless the person at that end allows his hand to move so as to absorb all of the energy, some of it will be reflected as a wave moving back toward the source.

(a) What is carried by a wave? _____

(b) How does energy transfer occur in the case of longitudinal waves? _____

_____

_____

Answers: (a) energy; (b) The person starting the wave puts energy in the spring, and it travels to the other end where it exerts a force through a distance on the person there. (The force is back and forth rather than up and down, but the principle is the same.)

10   THE DOPPLER EFFECT

If pebbles are regularly dropped at the same location into a pool of water, the wave pattern produced will look somewhat like that shown in figure 1 (below) after four pebbles are dropped. If the location of the drop is changed regularly toward the right so that the waves start from different places, a pattern like that of figure 2 will emerge. Moving the starting point even more rapidly produces the pattern of figure 3.

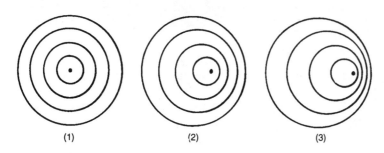

(1)   (2)   (3)

Now consider figure 3. Suppose one person is standing to the left of the place where the pebbles are being dropped (behind the moving pebble-dropper), and another person is to the right. Which person sees waves of longer wavelength approaching him? _____

_____

Answer: the one to the left

---

11   Since the speed of the waves is not affected by their wavelength, a person on the right of the source in figure 3 will see waves of greater frequency than the frequency at which the pebbles were dropped. And a person to the left of the source in figure 3 will see lower frequency waves.

(a) Does the apparent increase in frequency occur when the source is moving toward the observer or away? _____

(b) When the speed of the source is greater, is the frequency change greater or less? _____

Answers: (a) toward;   (b) greater

---

12   The effect we have been discussing is called the **Doppler effect**. It also occurs if the observer is moving and the source is stationary. As one moves toward the wave source, the apparent wave frequency is increased.

If the source continues to move faster until it is moving faster than the speed of the waves themselves, a pattern results such as that shown in the figure on the next page. Notice that along a "V" spreading from the source, the waves overlap and result in an extra-large wave. This is the **bow wave** that you see spreading from a fast-moving boat on a lake.

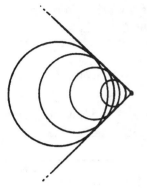

(a) How fast (relative to the wave speed) must a boat move to produce a bow wave? _____

(b) What is the name of the effect in which there is an apparent change in frequency due to the relative motion of the source and observer?

_____

(c) What happens if the observer is moving away from the source of waves? _____

Answers: (a) faster than the speed of the waves it produces;   (b) Doppler effect;   (c) The frequency observed will be less.

SELF-TEST

The questions below will test your understanding of this chapter. Use a separate sheet of paper for your diagrams or calculations. Compare your answers with the answers provided following the test.

1. Draw a simple wave and show what its wavelength is.

2. Locate the amplitude of the wave on the drawing for question 1. _____

3. You drop a rock in a swimming pool, producing a wave. It reaches the other side, 24 feet away, 4 seconds later. What is the speed of the water wave?

_____

4. If you push your hand in and out of the water regularly, you can produce a uniform pattern of waves. Suppose you dip your hand in and out 3 times a second and produce wavelengths of 0.6 meters. What is the speed of the waves?

_____

5. The speed of a particular wave is 30 meters/minute. The frequency of the wave is 2 cycles/second. What is its wavelength? _____

6. What is the period of the wave in question 5? _____

7. Distinguish between a transverse and a longitudinal wave. _____

_____

8. What is transferred as a wave travels along? _____

9. As a source of waves moves toward you, how does the apparent frequency of the waves compare to when the source was not moving? _____

_____

10. What happens to the apparent frequency as you move away from a source of waves? _____

11. Explain what conditions are necessary to form a bow wave and how this wave is formed. _____

_____

12. *Optional.* What is the period of a pendulum 4 feet long? _____

ANSWERS

If your answers do not agree with those given below, review the frames indicated in parentheses before you go on to the next chapter.

1. Either of the two distances labeled λ (or others this same length) is the wavelength. (frame 5)

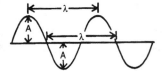

2. Either distance labeled A in the figure is the amplitude. (frame 5)

3. 6 ft/s (24 ft/4 s) (frame 5)

4. 1.8 m/s ($v = \lambda f = 0.6$ m·3/s = 1.8 m/s) (frame 6)

5. $\frac{1}{4}$ meter or 25 cm (The trick here is that the speed was given in meters per minute while frequency was expressed in cycles per second. 30 m/min = $\frac{1}{2}$ m/s.) (frame 6)

6. $\frac{1}{2}$ second (frame 3)

7. In a transverse wave the particles vibrate perpendicular to the direction of travel of the wave, while in a longitudinal wave they vibrate parallel to that direction. (frame 8)

8. energy (frame 9)

9. The apparent frequency is greater. (frames 10, 11)

10. The apparent frequency is less than when you were stationary. (frame 12)

11. The source of the waves must be moving faster than the waves themselves move. The waves then "stack up" along a "V" to form the bow wave. (frame 12)

12. 2.2 seconds

Solution: $t = 2\pi\sqrt{\dfrac{l}{g}}$

$$= 2 \cdot 3.14 \cdot \sqrt{\dfrac{4 \text{ ft}}{32 \text{ ft/s}^2}}$$
$$= 6.28 \cdot 0.354 \text{ s}$$
$$= 2.2 \text{ s} \qquad \text{(frame 2)}$$

11 Sound

Prerequisite: Chapter 10
Sound travels through a material in the form of longitudinal waves. Although we seldom observe directly the wave nature of sound, a consideration of this nature is necessary to understand the sound phenomena that surround us.

OBJECTIVES

After completing this chapter, you will be able to

- describe a sound wave in terms of air molecules;

- identify sound as a longitudinal wave;

- calculate the speed of sound in air at a given temperature;

- calculate the speed of sound using data concerning an echo;

- given either the wavelength or the frequency of a sound, calculate the other quantity;

- relate the speed of sound in a material to the elasticity and density of the material;

- relate intensity to the amplitude of a sound wave;

- calculate relative intensities of two sounds using the inverse-square relationship;

- explain the phenomenon of resonance as it relates to sound waves;

- identify some causes and results of refraction of sound waves;

- compare the speed of sound in air to its speed in other materials;

- identify the frequency range of audible sound;

- calculate the sound intensity in SI units given the sound's decibel level (optional).

1   SOUND WAVES

Part 1 of the figure is a representation of a sound wave. **Condensations** (compressions) and rarefactions follow one another through the material as the wave moves along. Since air is the material through which we normally experience the motion of sound, each dot here represents billions of molecules of air. At point X the molecules are in the center of a compression. As the wave moves along, that point will change to a rarefaction, then back to a compression, and so on.

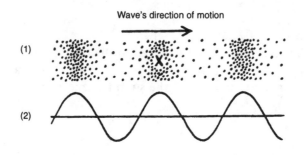

In part 2 of the figure the sound wave is represented graphically, with the height of the curve representing the pressure of the wave above or below normal pressure. Note that where there is a compression of the wave, the curve is at a maximum above center, indicating a high pressure. And where there is a rarefaction the curve is at a maximum below center, representing a minimum of pressure. A graph of the density of the air along the wave would be identical to this.

The fact that part 2 of the figure appears to represent a transverse wave can cause confusion. Sound is not a transverse wave and we are not calling it one. Remember that the curve is a graph of the pressure (or density), and is not meant to be a drawing of the wave itself.

(a) What corresponds to the highest point of amplitude, compression or rarefaction? _____

(b) Can you tell from this frame why sound doesn't travel in a vacuum? _____
Explain. _____

Answers: (a) compression; (b) Yes. In a vacuum, there is nothing to be compressed or rarefied.

2

The next figure represents the speaker of a radio or stereo. If you touch such a speaker (gently!) while it is producing sound, you will feel a vibration. It is this in-and-out motion of the speaker that sets up the sound wave in the air. In the figure the speaker cone has just moved to the right. This pushes the molecules of the air together in front of the speaker, causing a compression. Just as it does in a spring, the compression in the air moves away from the source. Meanwhile, the

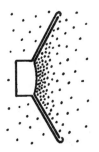

speaker cone is moving toward the left and producing a rarefaction in front of it. This rarefaction follows the compression through the air. Then the cone moves again toward the right, producing another compression, and the cycle continues. Thus, a sound wave moves away from the speaker through the air. The air itself is not moving away from the speaker; the molecules of air vibrate back and forth as the wave moves toward the right.

Your eardrum is somewhat similar to a speaker cone, but it is used as a receiver. When a compression hits the eardrum, the higher pressure causes the eardrum to move inward. Then the following rarefaction causes it to move outward. The vibrations are changed to electrical signals by the inner ear and then are transmitted to the brain for interpretation.

(a) How do molecules of air vibrate in a sound wave, perpendicular or parallel to the direction of the wave? _____

(b) What always follows a compression in a sound wave? _____

(c) What is transmitted in a sound wave, air molecules or energy? _____

Answers: (a) parallel; (b) rarefaction; (c) energy

3 THE SPEED OF SOUND

The speed of sound in air is relatively easy to measure. Suppose you stand in an open lot at a distance from the flat side of a large building and clap your hands. You will hear an echo when the sound is reflected from the building back to you. If you are 170 meters (about 550 feet) from the reflector, the echo will be heard about 1 second after the clap. Remembering that the sound travels to the building and back to you in the 1 second, calculate the speed of sound.

Answer: 340 m/s

4 The speed you calculated, 340 meters per second, is a representative speed of sound in normal air. It corresponds to about 750 miles per hour, $\frac{1}{5}$ mile per second, or 1100 feet per second. The speed of sound in dry air at a temperature of 0°C (32°F) is 330 m/s, but the speed depends upon the temperature and humidity of the air. For each Celsius degree rise above 0° the speed increases about 0.6 m/s.

The speed of sound in water is about 1490 m/s, more than four times greater than in air. In iron, the speed is about 5130 m/s. In general, the speed of sound is greater in more elastic materials and less in more dense materials. (Although iron is more dense than water, it is far more elastic, resulting in the much greater speed.)

(a) Seventy-two degrees Fahrenheit is about 22°C. What is the speed of sound at this temperature? _____

(b) How far will sound travel in 3 seconds in air at 22°C ? _____

(c) How far will sound travel in water in 3 seconds? _____

(d) Does sound travel faster in iron or in water? _____

Answers: (a) 343 m/s. Solution: 22·0.6 = 13; 13 + 330 = 343;   (b) 1030 m;   (c) 4470 m;
(d) iron

5   FREQUENCY AND WAVELENGTH RANGE

The human ear is sensitive to frequencies ranging from about 20 cycles/s to 20,000 cycles/s. Waves with frequencies greater than 20,000 cycles/s are said to be **ultrasonic waves**. Using the relationship between speed, frequency, and wavelength from the last chapter ($v = \lambda f$), calculate the range of wavelengths of sound in air. (Use 340 m/s as the speed of sound.)

(a) What is the wavelength of a 20-cycle/s sound? _____

(b) What is the wavelength of a 20,000-cycle/s sound? _____

Answers: (a) 17 m (340/20);   (b) 0.017 m, or 1.7 cm (less than an inch)
(See Chapter 10, frames 6 and 7, for a review of the equation used here.)

6   INTENSITY

The **intensity** of a sound wave is defined as the amount of sound power passing through a unit area. In Chapter 3, power was defined as energy divided by time. Therefore, intensity is the amount of energy passing through a square meter in 1 second. Watts/m^2 is the SI unit of intensity. The difference between the pressure (or density) in a compression and the pressure in a rarefaction is what determines the intensity of the wave.

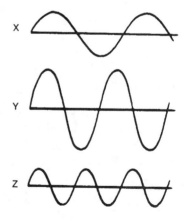

(a) On a graphical representation of sound waves, will the intensity be shown by the amplitude, the wavelength, or the frequency?

_____

(b) Which of the sound waves in the figure has a greater intensity? _____

Answers: (a) amplitude;   (b) Y

7 The actual amount of energy carried by a sound wave is small. It would take thousands of people shouting to produce enough sound energy to light a regular light bulb, even if all of the sound energy could be changed to electrical energy. The pressure difference between a compression and a rarefaction in air is only about 0.1 N/m² (about 10^{-5} lb/in.²) for conversational sound levels. (Recall that regular atmospheric pressure is about 10^5 N/m², or 15 lb/in.².)

Would a sound wave with a pressure difference of 10^{-4} lb/in.² be more or less intense than conversational sound level? _____

Answer: more (See Appendix I for powers-of-ten notation.)

8 THE DECIBEL SCALE

The metric unit of sound intensity is watts/m². While this unit corresponds to the system that is used in all fields of physics, intensity of sound has historically been expressed in decibels (abbreviated dB). The decibel scale is a compressed scale of measurement. In this system, the least intense sound that can be heard by a good ear has an intensity of 0 decibels. A sound carrying ten times as much energy is rated as 10 decibels, a sound with 100 times as much energy as the 0-dB sound is rated at 20 dB, and 1000 times as much energy at 30 dB.

On the decibel scale a whisper measures about 20 dB, and ordinary conversation is about 60 dB. Sound becomes painful at about 120 dB, the "threshold of pain." At a distance of about 100 yards, a rocket engine has an intensity of about 180 dB, well above the threshold of pain.

(a) A sound of how many decibels has 10,000 times the energy of a 0-dB sound?

_____

(b) Is this above the threshold of pain? _____

(c) Rock concerts have been measured at 125 dB. Is this above the threshold of pain? _____

Answers: (a) 40 dB; (b) no, not nearly; (c) yes (Prolonged sound at this level causes permanent hearing damage.)

9 INTENSITY CALCULATIONS (OPTIONAL)

A 0-decibel sound has an intensity of 10^{-12} watt/m². A 10-dB sound thus has an intensity of 10^{-11} watt/m². Recall that an increase of 10 dB corresponds to a tenfold energy increase.

(a) What is the intensity (in SI units) of a 20-dB sound? _____

(b) How much more energy than 0 dB does 20 dB carry? _____

(c) How many times more intense is an 80-dB sound than a 50-dB sound?

_____

(d) What is the intensity in watt/m² of a 70-dB sound?   _____

Answers: (a) 10^{-10} watt/m²;   (b) 100 times;   (c) 1000 times (because the difference in decibels is 30);   (d) 10^{-5} watt/m² (An increase of 70 dB multiplies the intensity by 10^7 times: $10^7 \cdot 10^{-12} = 10^{-5}$.)

10   DECREASE OF INTENSITY WITH DISTANCE

When a pebble is dropped into a calm lake, a ripple spreads over the surface of the water in an ever-enlarging circle. Consider the energy carried by the wave. As the circle gets larger and larger, the energy becomes more spread out, so that by the time the ripple arrives at a distant shore, the ripple is barely perceptible. Except for the energy that is changed into heat due to the stirring of the water, the original amount of energy is still in the ripple, but it has spread over such a large circle that the energy contained in a short section of the circle is very little.

When you ring a bell, the sound waves produced spread out in much the same way that the water wave did, except that the spreading is now in three dimensions rather than in just two. Instead of an ever-enlarging circle, we have an ever-enlarging sphere. The area of a sphere is given by the formula:

$$A = 4\pi r^2$$

where r is the radius of the sphere. Intensity is defined as the power passing through a unit area. Since the area of a sphere depends upon the square of the radius (r^2), the energy carried by the wave spreads out inversely proportional to the square of the radius. Thus, intensity, I, depends upon the reciprocal of the radius squared, or:

$$I \sim \frac{1}{r^2}$$

Assume that the intensity of sound 10 meters from a certain source is 10^{-6} W/m².

(a) Using the formula, how intense will the sound be at 20 meters (twice as far away as 10 meters)?   _____

(b) How intense will it be at 30 meters?   _____

Answers: (a) $0.25 \cdot 10^{-6}$ W/m² (It is $\frac{1}{4}$ as intense since it is twice as far away.); (b) $0.11 \cdot 10^{-6}$ W/m²

Actually, we have neglected the fact that some sound energy is absorbed by the air. Thus, the actual sound intensity is reduced even more quickly than indicated in frame 10. The inverse-square effect due to the spreading of the energy is the major cause of the decrease in intensity, however.

One way to reduce the effect of the spreading is to "funnel" the sound in a particular direction by using a device such as a megaphone. This causes most of the energy to be sent in the desired direction. Three sources of sound are shown in the figure. Each emits the same power.

Which produces the loudest sound at K? _____

Answer: source Z

12 RESONANCE

Every object has a natural frequency at which it will vibrate. Kick the desk in front of you and you will probably hear a low-frequency sound. If you drop a spoon, it will vibrate with a higher frequency. You may have had the experience of hearing a table or cabinet vibrate when a speaker (playing loudly) is placed on or near it. This is an example of **resonance**. As a sound wave travels through the air it carries energy with it. This energy is what causes our eardrums to vibrate when struck by the wave. In the same manner, a larger, less sensitive object also bends slightly with each compression and rarefaction that hits it. Normally this deformation is so slight that it is not noticeable, but if the frequency of the sound wave just matches the natural frequency of the object struck, the object will pick up the vibration—it will **resonate**. It is said that some singers can shatter a crystal goblet with their voice. If so, they are producing a sound that matches the natural vibrating frequency of the crystal.

(a) What is it about a loud sound that enables it to break the goblet when a quieter sound won't? _____

(b) Why will the goblet break due to sound of one frequency but not sound of a different frequency? _____

Answers: (a) The louder sound carries more energy. (Saying that the louder sound has a greater amplitude is an equivalent answer.);   (b) The frequency of the sound must match the natural frequency of the goblet.

13 REFRACTION

We said earlier that sound travels faster in warm air than in cold. Consider what happens when air near the ground is cooler than air above the ground. As sound waves travel away from a source, those going up into the warmer air start to travel faster. This results in the waves being bent from their original directions, as shown in the figure. The effect is that the waves tend to be bent back down toward the ground. Such a bending of waves is called **refraction**. Refraction occurs whenever waves pass at an angle into an area where they travel at a different speed.

Notice in the figure below that the same number of waves strike the listener each second as leave the source each second, so refraction does not affect the frequency that is heard.

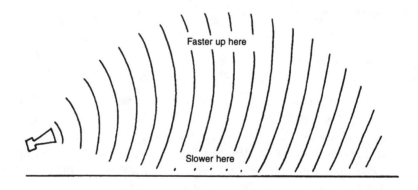

(a) Indicate which of these might be a factor in refraction:

_____ air temperature

_____ intensity

_____ the material the wave passes through

(b) Would you expect refraction to have an effect on the wavelength or on the frequency of the sound? _____

(c) What happens to the direction of a wave when it is refracted? _____

_____

(d) What would happen to a sound wave if the air near the ground were warmer than above (so that the sound traveled faster near the ground)? _____

_____

Answers: (a) air temperature and material; (b) wavelength; (c) It changes. (d) The sound would bend upward.

SELF-TEST

1. Are sound waves transverse or longitudinal?_____

2. How do the molecules in a compression differ from those in a rarefaction?

   _____

3. What is the speed of sound in air at a temperature of 18°C? _____

4. What would you predict concerning the speed of sound in a metal that has the same density as iron but is less elastic than iron? _____

5. What are the lower and upper limits of frequency to which the human ear is sensitive? _____

6. A frequency of 500 cycles per second in air (at 0°C) has what wavelength?

   _____

7. A 50-dB sound carries how many times as much energy as a 0-dB sound?

   _____

8. Suppose the intensity at 50 feet from a siren is 10^{-5} watt/m^2. What is the intensity at a distance of 100 feet? _____   At 200 feet? _____

   _____

9. If one moves twice as close to a source of sound, how does the intensity reaching him change? (Assume that there is negligible reflection of sound from surroundings.) _____

10. What must be different about how sound travels in two materials in order for refraction to occur when sound goes from one material to the other?

    _____

11. *Optional.* What is the intensity in decibels of a sound that has an intensity of 10^{-4} watt/m^2? _____

12. *Optional.* What is the intensity (in watts/m^2) of a sound of 50 dB? _____

    _____

13. *Optional.* How many times more intense is a 70-dB sound than a 40-dB sound?

    _____

ANSWERS

1. longitudinal (frame 1)

2. They are closer together (and therefore under more pressure) in a compression. (frames 1, 2)

3. 341 m/s. Solution: A rise of 18 C° above the zero temperature causes 0.6·18 m/s rise in speed over the speed of 330 m/s at 0°C. (frame 4)

4. It would be less. (frame 4)

5. 20 and 20,000 cycles per second (frame 5)

6. 0.66 meter (This is obtained by dividing the speed—330 m/s—by the frequency.) (frame 5)

7. 100,000 times (frame 8)

8. $0.25 \cdot 10^{-5}$ watt/m^2 (because at twice the distance the intensity is reduced to $\frac{1}{4}$ as much); $0.0625 \cdot 10^{-5}$ watt/m^2 (because it is four times as far away, which makes the multiplying factor $(1/4)^2$). (frame 10)

9. It gets four times as great. (frame 10)

10. The speed of sound must be different in the two materials. (frame 13)

11. 80 dB (10^{-4} watt/m^2 is 10^{-12} watt/m$^2 \cdot 10^8$. Thus, the 0-dB level sound has been increased by a factor of 10^8. This corresponds to 80 dB.) (frame 9)

12. 10^{-7} watt/m^2 (50 dB is 10^5 times more intense than 10^{-12} watt/m^2) (frame 9)

13. 1000 times more intense ($70 - 40 = 30$. Thus, the intensity has been increased by a factor of 10^3.) (frame 9)

12 Diffraction, Interference, and Music

Prerequisites: Chapters 10 and 11

In this chapter we will look further into the wave nature of sound and consider phenomena that require us to delve into what happens when two sound waves overlap.

OBJECTIVES

After completing this chapter, you will be able to

- relate loudness to intensity;
- relate pitch to frequency;
- differentiate between refraction and diffraction;
- describe the cause of sound wave interference;
- given a drawing of a standing wave, specify the number of wavelengths;
- differentiate between constructive and destructive interference;
- describe the conditions necessary for beats to be produced;
- determine the beat frequency of two sound sources of given frequencies;
- specify what determines the quality of sound;
- relate a sound's overtones to its fundamental tone in terms of frequency;
- differentiate between standing and traveling waves;
- identify nodes and antinodes in a standing wave;
- relate the Doppler effect to sound waves;
- describe the conditions necessary to produce a sonic boom and explain what causes it.

1 LOUDNESS

In our everyday language we may use the term loudness to mean intensity. Intensity, however, refers to a measurable, physical attribute of a sound wave, while loudness concerns a physiological sensation caused by a sound wave. You may perceive the same sound to be louder than I do, but intensity is measurable and not subject to individual interpretation. In general, however, the more intense the sound, the louder it seems.

For one sound to seem twice as loud as another, its intensity must not be twice as much—but eight to ten times as much. Because of this logarithmic effect of loudness, our ears are sensitive to a very great range of sound intensities. We detect sound energies as weak as 10^{-12} watt/m^2, and we can hear without pain sounds with intensities up to about 1 watt/m^2 (although extended exposure to intensities above 10^{-3} watt/m^2—90 dB—can cause permanent hearing loss).

Suppose you know that the sound of a whisper is 20 dB (so that it has intensity of 10^{-10} watt/m^2). You hear another sound that seems to be twice as loud as a whisper. What would you guess its intensity to be: twice 10^{-10}, one-half 10^{-10}, or ten times 10^{-10}?

Answer: $10 \cdot 10^{-10}$ (which is 10^{-9})

The above method of determining loudness is approximate, for in addition to intensity, the frequency of a sound also has an effect on the perceived loudness.

2 DIFFRACTION OF SOUND

Diffraction is defined as the spreading out of a wave as it goes around a corner. It is obvious that sound waves do spread around corners, for you can hear a person in the next room even though he may be out of sight. (Some sound actually passes through the wall, but most of the sound you hear comes through the door opening.)

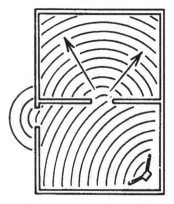

Notice that the bending that occurs in diffraction differs from bending that occurs in refraction. When waves are refracted they no longer continue in the same direction, but in the case of diffraction only part of the wave goes in a new direction. Consider sound going through a door from one room to another. As shown in the figure, the sound continues straight into the second room as well as bending around the door opening.

Label the drawings below as diffraction or refraction.

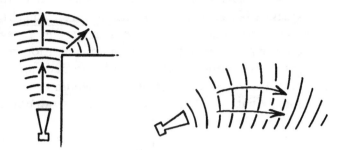

Answer: The left one is diffraction, and the right one is refraction (see Chapter 11, frame 13).

3   INTERFERENCE

Consider a person listening to a sound of constant frequency coming from two speakers as shown in the figure here. Each speaker is placed at the same distance from the listener, and each is connected to the amplifier so that when one is producing a condensation the other is also. When these conditions are met, the person is struck by condensations from both speakers at the same time, then by rarefactions from both speakers. The waves are said to be striking him **in phase**. Both speakers cause an increase in pressure on the listener's eardrum at the same time and a decrease in pressure at the same time.

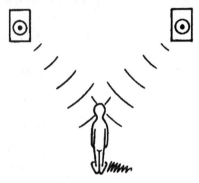

(a) When two sound waves strike a detector in phase, is the combined effect of a rarefaction more, less, or the same pressure than it would be if only a single source were present? _____

(b) How would you expect in-phase sound to affect the amplitude of a sound wave?

_____

(c) What effect on the loudness would you expect due to two sound waves being in phase? _____

Answers: (a) less pressure (and a condensation would have greater pressure due to the combined waves); (b) The sound wave that results would have greater amplitude. (c) The sound would be louder.

4   Now consider the case shown in the figure in this frame. Here one speaker is located exactly one-half wavelength further from the listener than the other speaker. Now when a condensation from the left speaker reaches his ear, a rarefaction is reaching it from the right speaker. Likewise, when a rarefaction reaches the ear from the left speaker, a condensation reaches it from the other. In each case a condensation alone would produce an increase in pressure over normal pressure while a rarefaction alone would produce an equivalent decrease in pressure. What would you expect as the overall effect? _____

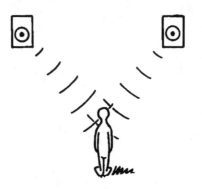

Answer: They would cancel each other out.

5   When two sound waves are in phase, so that there is increased intensity and loudness of sound, we say that we have **constructive interference**. When the waves cancel one another out, resulting (in the ideal case) in no sound being heard by the listener, we call it **destructive interference**. (In practice, complete cancellation is difficult to achieve, because it would require perfect spacing, identical sound sources, and no reflection from surrounding surfaces.)

(a) What is interference in sound waves? _____

_____

(b) How does constructive interference differ from destructive interference?

_____

Answers: (a) One wave affects or changes another when the two overlap.   (b) Constructive interference increases intensity while destructive interference decreases it.

6   Recall that sound waves may be represented by pressure graphs. Figure A on the next page represents the sound wave from some source, and B represents another sound wave that is located at the same place as wave A. Figure C shows the combined effect—the waves added together.

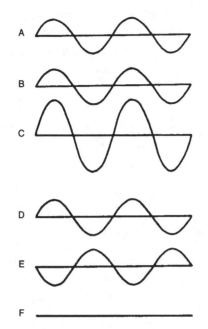

(a) Does Figure C represent constructive or destructive interference? _____

_____

(b) The lower three graphs represent two more waves (D and E) and their total (F). Are D and E in phase or out of phase? _____

(c) Do D and E interfere constructively or destructively? _____

Answers: (a) constructive;   (b) out of phase;   (c) destructively

7   BEATS

The figure in this frame represents two side-by-side speakers emitting sound waves that differ slightly in frequency. The speaker at top has the longer wavelength and therefore the lesser frequency. The waves are shown traveling beside one another, but in actual practice they would overlap, causing interference. Where the two waves meet with condensation on condensation, constructive interference results, but where condensation lands on rarefaction, destructive interference results.

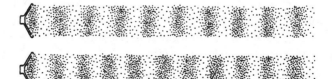

In waves A and B below, the two waves are shown in graphical form. The resultant wave is then shown as wave C. Note that there are areas of wave C where the amplitude is greater than the amplitude of either wave alone, but at other places the waves are canceled out.

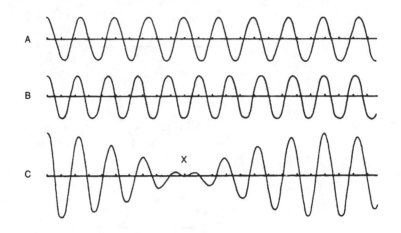

Now imagine this resultant wave moving toward a listener located at the right (but not shown in the figure). The listener hears sound when the large-amplitude waves hit her ear, but the sound dies down when the points of destructive interference arrive. The effect that is heard is a repeated changing of the loudness of sound. This effect of regular changing loudness is called **beats**. Beats can be heard whenever two sounds of nearly the same frequency are sounded together.

(a) Is point X in wave C a place of constructive or destructive interference?

_____

(b) Is the phenomenon of beats the result of constructive interference, destructive interference, or both? _____

Answers: (a) destructive;   (b) both

8

The **beat frequency** is the frequency with which the cycle of loud and not-so-loud sound occurs. This beat frequency is equal to the difference between the frequencies of the two sounds that produce the beats. Thus, if one source vibrates with a frequency of 500/s and the other vibrates at 503/s, the beat frequency is 3/s. A piano tuner uses a sound source of known frequency (normally a tuning fork) and adjusts the appropriate piano string until the beats between the two are eliminated.

If the frequencies of two sound sources are 420/s and 430/s, what beat frequency would be heard when the two sounds are produced together? _____

Answer: 10/s

9 ┃ PITCH

Pitch is the subjective impression that distinguishes sounds by differences in tone. We hear a higher pitch when we strike a key at the right end of a piano keyboard than when we strike one at the left. Pitch depends primarily on frequency, a higher frequency sound generally being heard as a higher pitch. But while frequency is a physical, measurable quantity, the pitch of a sound is a physiological sensation and is not easily measured.

(a) Is pitch more like loudness or intensity in terms of measurability? _____

_____

(b) Would a sound with a frequency of 500/s or 400/s be perceived as a higher pitch? _____

Answers: (a) loudness;   (b) 500/s

10 ┃ QUALITY

Thus far we have considered only sounds that can be presented graphically as smooth, regular waves.* In actual practice, few sounds are so "pure." A tuning fork, when struck correctly, produces a pure tone, and a whistle made by pursing the lips is often a pure tone, but most sounds we hear are combinations of a number of frequencies. The graphs at left show what the same note played on two

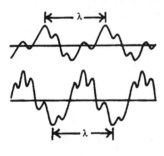

different musical instruments may look like. Note that the wavelength—the distance between complete repeating cycles—is the same for both. (The wavelength is measured for entire repeating cycles. Don't be distracted by the variations.) Since the basic wavelengths of the two sounds are the same, their fundamental frequency is the same. However, we can tell the sounds apart when we hear them. Though we would say the waves have the same frequency, our ears pick up a difference in the sounds. We say the two sounds differ in **quality**.

(a) How would you expect the two sounds represented by the graphs to compare in pitch? _____

(b) The two curves do differ in appearance. Would you expect them to sound identical? _____

Answers: (a) the same pitch;   (b) no

The quality of the sounds is what enables you to tell a piano from a guitar.

---

*Such regular waves are usually **sine waves**, so called because they can be expressed mathematically by the trigonometry function of the same name.

11 The quality of a tone is determined by the number and relative intensities of its overtones. To consider how these frequencies are produced, we will consider strings that are undergoing vibration. Such vibration can produce sound waves. When a string is fixed at its two ends it can be made to vibrate in a number of different ways.

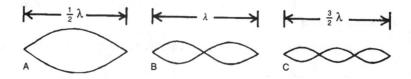

Part A of the figure shows the simplest vibration. The drawing is intended to illustrate a string that is vibrating up and down, but only the two extreme positions of its vibration are shown. This pattern of vibration is called the **fundamental mode of vibration.** If you ever tied one end of a long spring to something and shook the other end, you know that you can make the spring vibrate as shown in part A of the figure, and that if you shake your end faster you can produce a vibration such as in B. To get a vibration mode like B, you vibrate the spring with twice the frequency as you did to produce pattern A (which represents the lowest vibrating frequency). Figure C shows three segments in the spring. Its frequency of vibration is three times that of A.

If a spring is vibrating in the mode represented below, how does its frequency of vibration compare to its lowest possible vibrating frequency? _____

Answer: It has five times the frequency.

12 If the vibration frequency of a string is greater than 20 cycles per second, a sound can be heard, for the frequency of the wave is then within the range of hearing. The sound produced by the vibration shown in part B of the previous figure has twice the frequency of that in A. In practice, however, a string on a musical instrument vibrates in a number of different patterns at the same time. The lowest frequency of vibration, the fundamental, determines the pitch of the sound we hear. The next highest frequency, produced by the vibration of part B, is called the **first overtone.** The third mode shown (C) is the second overtone and has a frequency three times as great as the fundamental. The resulting sound wave that we hear is thus the product of interference of a number of waves.

(a) What determines the quality of a tone? _____

_____

(b) How does the frequency of the third overtone compare to that of the fundamental? _____

Answers: (a) the number and relative intensity of overtones; (b) It is four times as great.

13 STANDING WAVES IN A STRING

When a string is vibrated as in the large figure of frame 11, instead of waves
appearing to move down the string and back, the waves appear to be standing in
one place. This type of vibration is appropriately called a **standing wave**. In part A
the string was vibrating in its fundamental mode, and the distance from one end
of the string to the other was half the wavelength of the transverse wave in the
string. In B, the distance between ends was one wavelength. Since the speed of the
waves in the string does not depend upon the wavelength, we can conclude that
the frequency in B is twice as much as in A.

In C, the distance between ends was $1\frac{1}{2}$ wavelengths and the frequency was
three times as much as in the fundamental vibration. In general, when a string is
fixed at both ends, one can set up standing waves in it of $\frac{1}{2}$, 1, $1\frac{1}{2}$, 2, $2\frac{1}{2}$, ...
wavelengths. This is usually written in equation form as:

$$d = \frac{n\lambda}{2} \qquad n = 1, 2, 3, 4, \ldots$$

where d is the length of the string and λ is the wavelength. (An equation such as
this is really an infinite number of equations, for there is an equation for each
value of n.)

(a) How many wavelengths long is the string when $n = 5$? _____

(b) How does a standing wave differ from a moving one? _____

_____

Answers: (a) $2\frac{1}{2}$ (b) The waves just vibrate back and forth; they don't travel along the
string.

14

The next figure represents a standing wave $2\frac{1}{2}$ wavelengths long. The points
where the string is not in motion, points A, C, and E for example, are called
nodes. The points midway between the nodes, where the vibration is at its maximum, are called **antinodes**.

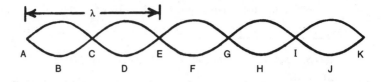

(a) Consecutive antinodes are separated by how many wavelengths? _____

(b) What is the distance between a node and the closest antinode? _____

_____

Answers: (a) $\frac{1}{2}$;   (b) $\frac{1}{4}$ wavelength

15   The standing waves shown thus far all have had nodes at their ends. Although it is not easy to set up in practice, one can imagine a standing wave in a string in which the qualifying conditions are that it has a node at one end and an antinode at the other. Such a wave is shown below.

(a) The length of the string here is how many wavelengths? _____

(b) How many wavelengths long is the string when it is vibrating at its lowest possible frequency under these qualifying conditions? _____

(c) The equation for the list of possible wavelengths for this type is:

$$d = (2n - 1) \cdot \frac{\lambda}{4} \qquad n = 1, 2, 3, \ldots$$

What value does this give for d when $n = 1$? _____

Answers: (a) $1\frac{1}{4}$;   (b) $\frac{1}{4}$ (A node would be at one end and an antinode at the other, with no other nodes or antinodes between.);   (c) $\frac{1}{4}\lambda$

(You might substitute some values of n in the equation to satisfy yourself that it works. For $n = 3$ we get the situation in the figure.)

16   STANDING WAVES IN A GAS

The reason that we have spent so much time on standing waves in a string is that they are much more easily visualized than the longitudinal standing waves of sound. Such waves occur when sound of the right frequency travels down a tube and is reflected back toward the open end. Although the waves in this case are longitudinal, we represent them graphically with drawings that look just like those of the transverse standing waves in a string.

A standing wave can be set up in a tube that has one end closed if there is an antinode at the closed end and a node at the open end.* Thus, the length of the tube will be $\frac{1}{4}$ wavelength for the lowest frequency wave. Part A of the figure here represents such a wave. The next highest frequency that will cause a standing wave in this tube is shown in B.

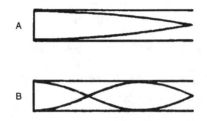

(a) How many wavelength(s) long is the wave in B? _____

(b) Does wave B represent the fundamental, first overtone, or second overtone?

_____

(c) How many nodes and antinodes are shown in B? _____

Answers: (a) $\frac{3}{4}$; (b) first overtone; (c) two of each

17 A longer tube produces a lower fundamental frequency, because the wavelength is longer. (You can use the equation $v = \lambda \cdot f$ to see this.) This is the principle of wind instruments and is carried to the ultimate in the pipe organ, where the tubes that produce high frequencies are only a few inches long, but the tubes for low frequencies are a number of feet long.

One may also set up standing waves in a tube that is open at both ends, because some sound will be reflected back from the open end. In this case a node will be at both ends. This situation is similar to that of a string that is held at both ends.

---

*If you consult other texts, you'll find that some reverse the positions of node and antinode. The wave they are considering represents the displacement of molecules in the wave. We are representing the pressure.

(a) At which length will the fundamental frequency occur for a tube open at both ends? _____

(b) How does this compare to the closed end pipe? _____

_____

Answers: (a) $\frac{1}{2}$ wavelength (from one node to the next);   (b) The length of a closed pipe is $\frac{1}{4}$ wavelength of the fundamental tone.

18   THE DOPPLER EFFECT IN SOUND

In Chapter 10 (frame 10) we saw that when a source of sound moves away from the observer, the frequency at which the waves pass the observer is less than when the source is stationary. As the source moves toward the observer, the frequency is greater. This phenomenon also occurs in the case of sound. Since a high frequency corresponds to a high pitch, and a low frequency to a low pitch, you can see that the Doppler effect explains the apparent change in pitch of the sound of a race car as it rushes by.

It is important to see what the Doppler effect is *not* saying. First, it is not concerned with the loudness of the sound. Sure, the sound becomes louder as the car approaches, but this is nothing more than the inverse square effect discussed in Chapter 11 (frame 10). Also, the Doppler effect does not say that the frequency increases as the car comes closer. It says that the frequency is *constantly* higher all of the time as the car approaches, and constantly lower as the car recedes.

(a) What explains the higher pitch of a police car siren when it is approaching you? _____

(b) What explains the fact that a jackhammer sounds louder when you are next to it than when you are far away? _____

Answers: (a) Doppler effect;   (b) inverse square law

19   THE SONIC BOOM

When a wave source on water moves at a speed greater than the speed of the water waves, a bow wave is produced. We can see now that the bow wave is in fact the result of constructive interference of waves as the peaks pile one on another. Again, a similar phenomenon occurs in the case of sound.

The next figure is similar to the figure in Chapter 10 (frame 12), but here the lines represent condensations of a sound wave. Notice that they pile up just as did the water waves. But since the sound waves spread not only in two dimensions, but up and down as well, the figure formed is not simply a "V," but a cone. A

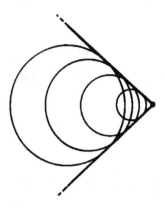

cone of highly compressed air spreads from any object that is moving faster than the speed of sound. When this compression reaches someone, he or she experiences a sudden, loud sound, a **sonic boom**. The pressure amplitude of sonic booms may be sufficient to break windows or crack plaster.

Note that a sonic boom is produced all the time that a plane is flying at a speed greater than the speed of sound, and *not* just at the instant that the plane goes from slower than the speed of sound to faster than the speed of sound (called "breaking the sound barrier" because in order to do this, the plane must push through the wall of high pressure in front of it).

(a) How fast must an object go to create a sonic boom? _____

_____

(b) Will a sonic boom be created by an object moving at a speed of 2000 ft/s?

_____

(c) Explain how a sonic boom can cause actual damage. _____

_____

Answers: (a) faster than the speed of sound (340 m/s, or 1100 ft/s) (Chapter 11, frame 4); (b) yes;   (c) The sudden pressure change can cause damage.

SELF-TEST

1. How does loudness differ from intensity of sound? _____

   _____

   _____

2. What is meant when it is said that two sound waves are "in phase"? _____

   _____

3. What is the difference between constructive and destructive interference?

   _____

   _____

   _____

4. What conditions are necessary to produce beats? _____

_____

5. If two tuning forks—one 723 cycles/s and the other 727/s—are sounded together, what beat frequency is heard? _____

6. What is the difference between pitch and frequency? _____

_____

_____

7. What determines the quality of a note? _____

_____

8. How are we able to distinguish between a note played on a flute and the same note played on a clarinet? _____

9. If a string is held at both ends, how would it appear when vibrating at its fundamental frequency? _____

_____

10. The figure represents a string in vibration. Its frequency of vibration is how many times as great as its fundamental frequency (when held the same way)? _____ What overtone does this vibration represent? _____

_____

11. What is a common phenomenon that is explained by the Doppler effect in sound? _____

12. What conditions are necessary to produce a sonic boom? _____

_____

13. How many wavelengths long is the string in the figure of question 10? _____

14. A long whip can be made to "crack" by making its end to move at a speed greater than the speed of sound. What causes the "crack" or "pop"?

_____

ANSWERS

1. Loudness is the physiological sensation while intensity is a direct measure of the power per square meter of the wave. (frame 1)

2. "In phase" means that when one wave is at a condensation, the other is also. (frame 3)

3. Constructive interference is the combination of two waves such that the result is of greater intensity than either wave alone, and destructive interference results in the total wave having less intensity than either wave alone. (frame 5)

4. Two waves must be produced at nearly the same frequency. (frame 7)

5. four cycles per second (frame 8)

6. Pitch is the physiological sensation while frequency is a physical characteristic of the sound wave. (frame 9)

7. the number and relative intensities of overtones (frame 11)

8. They differ in quality. (frame 10)

9. It would have nodes only at the ends, with maximum vibration at the center. (frame 11)

10. three; the second overtone (frames 11, 12)

11. The noise produced by a passing race car, police car, train, etc. (frame 18)

12. The source of sound must be moving faster than the speed of sound. (frame 19)

13. $1\frac{1}{2}$ (frames 13, 14)

14. sonic boom (frame 19)

13 Static Electricity

Prerequisite: Chapter 5
The phenomenon of static electricity is a familiar one. Static electricity causes the shock when we slide across a car seat and touch the door handle, static electricity makes a sweater cling when we take it off, and static electricity causes the beautiful but destructive lightning bolts.

OBJECTIVES

After completing this chapter, you will be able to

- compare the electrostatic forces between two charges when the distance between them is changed;

- determine the relative strengths of electric fields using Coulomb's law;

- specify the units of electric charge and electric field strength;

- determine the electrostatic force on a given charge in a given electric field;

- interpret a diagram of electric field lines in terms of direction and relative strength;

- give examples of charging by friction and charging by induction;

- differentiate between electron action in conductors and insulators;

- describe the effect of a charged object on an electroscope;

- describe what happens to the atomic charges of the objects when two objects are charged by friction;

- calculate the force between two charges, given the strength of the charges and the distance between them (optional);

- calculate the strength of an electric field at a given distance from a given charge (optional).

1   COULOMB'S LAW

The entire phenomenon of electricity is caused by the action of the charged particles that make up atoms: protons (which are positive) and electrons (which are negative). The basic force law of electrostatics is that opposite charges attract one another and similar charges repel one another. A negatively charged object attracts a positively charged object. Two positive objects or two negative objects, on the other hand, repel one another. Consider an atom that has lost one electron.

(a) Will it be positively or negatively charged? _____

(b) Will it attract or repel an atom that has gained an extra electron? _____

Answers: (a) positively (It has lost a negative charge.);   (b) attract

2

The strength of the attraction or repulsion force is determined by Coulomb's law:

$$F \sim \frac{Q_1 \cdot Q_2}{d^2} *$$

where Q_1 and Q_2 are the strengths of the two charges and d is the distance between them.

This equation is very similar to the equation governing gravitational forces (Chapter 4, frame 1). Both force laws state that the force is inversely proportional to the square of the distance between the objects.

(a) If two charges are initially 1 meter apart and are then separated to a distance of 2 meters, the force is reduced to what fraction of its original value? _____

(b) If the distance is made 3 meters, the force is what fraction of the original?

_____

Answers: (a) $\frac{1}{4}$;   (b) $\frac{1}{9}$

3

In Chapter 1 it was stated that all quantities can be measured in terms of mass, distance, time, and electrical charge. Electric charge is measured in coulombs, a unit independent of the units of mass, distance, and time. A coulomb is essentially a certain number of electrons (in the case of negative charge) or protons (in the case of positive charge). A coulomb is the amount of charge on $6.2425 \cdot 10^{18}$ electrons, or, in conventional notation, 6,242,500,000,000,000,000 electrons. That is a lot of electrons! (Don't memorize the number, by the way.)

---

*The symbol "~" means "is proportional to."

(a) What element(s) in the expression $F \sim \dfrac{Q_1 \cdot Q_2}{d^2}$ would be given in coulombs?

_____

(b) What element(s) would be given in meters? _____

Answers: (a) Q_1 and Q_2; (b) d

---

4 CALCULATIONS WITH COULOMB'S LAW (OPTIONAL)

The SI unit of force is the newton, the unit of charge is the coulomb, and the unit of distance is the meter. If 1 coulomb of charge is located 1 meter away from another coulomb of charge, the electrostatic force between the two is tremendous: $9 \cdot 10^9$ newtons. That number, then, is the constant that must be used to change the statement of proportionality (in the previous frames) into an equation. Thus, Coulomb's law becomes:

$$F = 9 \cdot 10^9 \cdot \frac{Q_1 \cdot Q_2}{d^2}$$

What are the SI units in this equation?

F: _____; Q_1 and Q_2: _____; d: _____

Answer: newtons; coulombs; meters

5

A force of 9 billion newtons is not seen in normal cases of electrostatic force, since a coulomb is a far larger charge than is generally encountered. A normal charge may be a few microcoulombs, where

$$10^6 \text{ microcoulombs} = 1 \text{ coulomb}$$

Suppose you charge two small combs with 0.2 microcoulombs each by combing your hair.

(a) How much charge, in coulombs, is on each comb? _____

(b) What force do the combs exert on each other when they are 10 centimeters apart? _____

Answers: (a) $2 \cdot 10^{-7}$ coulombs;

(b) 0.036 newtons. Solution: $F = 9 \cdot 10^9 \cdot \dfrac{(2 \cdot 10^{-7})(2 \cdot 10^{-7})}{(0.1)^2}$

$$= 36 \cdot 10^{-3}$$
$$= 0.036$$

(Since a newton is about $\frac{1}{4}$ pound, this comes to about 0.01 pound, or somewhat more than 0.1 ounce.)

6 THE ELECTRIC FIELD

Suppose we have a small charged object at a particular location. If another charged object were to be brought near that object, a force would be exerted on it, as described by Coulomb's law. It is convenient to say that the original charged object is surrounded by an "electric field." The figure shows an isolated positive charge. The arrows that radiate outward from the charge indicate the direction of the force that will be experienced by a positively charged object that is brought into that area. These are sometimes called "lines of force."

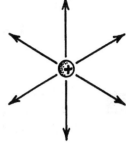

(a) Does the concentration of lines of force get more or less as they extend outward? _____

(b) Does the force on a positively charged object become greater or less as the object gets farther from the charge in the figure? _____

Answers: (a) less;   (b) less

The figure here shows an isolated negative charge. Here the arrows on the lines of force point toward the charge. Again, they show the force a positive charge would experience in that area.

Where would the force on a positive charge be greatest?

_____

Answer: near the object

As we have seen, lines of force are closer together where the force is greater. At greater distances from a charge, the lines separate—and the force diminishes. The figure in this frame shows two charges, one positive and one negative, and the force field surrounding them.

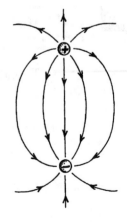

(a) Where is the origin of each force line?

_____

(b) Where is the destination of each force line?

_____

(c) Where does the density of lines indicate that the force is strongest?

_____

Answers: (a) at the positive charge; (b) at the negative charge; (c) near each charge

9

The strength of an electric field at a certain point is equal to the strength of the force on a charge of 1 coulomb placed at that point. The unit of electric field strength, therefore, is "newtons/coulomb." Once we know the strength of an electric field in newtons/coulomb, we no longer need to know the configuration of charges that caused that field in order to know what force the field will exert on a certain charge. For example, suppose we bring a charge of 0.05 coulombs to a point in an electric field where the electric field strength is 250 newtons/coulomb. How do we find the amount of electrostatic force that is experienced by the 0.05-coulomb charge? From the fact that the strength of the field at the point is 250 newtons/coulomb, we know that a force of 250 newtons will be exerted on each coulomb of charge that may be brought to that point. The force on our 0.05-coulomb charge is therefore 0.05 times 250, or 12.5 newtons. Notice that we did not need to know where the charge (or charges) was located that caused the electric field.

How much electrostatic force would be experienced by a charge of 0.02 coulombs placed where the electric field has a value of 30 newtons/coulomb?

_____

Answer: 0.6 newtons

10 CALCULATING ELECTRIC FIELDS USING COULOMB'S LAW (OPTIONAL)

To calculate the electric field near a charge, you assume that a charge of 1 coulomb is located at the given distance from the given charge, and you use Coulomb's law to calculate the force on this assumed 1-coulomb charge. Now use the formula $F = 9 \cdot 10^9 \cdot (Q_1 \cdot Q_2)/d^2$ to find the strength of an electric field at a distance of 0.5 meters from a positive charge of 3 microcoulombs. _____

_____

Answer: $1.08 \cdot 10^5$ newtons. Solution: $F = 9 \cdot 10^9 \cdot \dfrac{(3 \cdot 10^{-6})(1)}{(0.5)^2}$

(This answer means that there are 108,000 newtons of force on each coulomb at that point in the field. We say that the field strength is 108,000 newtons/coulomb.)

11 CHARGING BY FRICTION: THE TRIBOELECTRIC SERIES

We normally see the phenomenon of static electricity when two objects are rubbed together. This method is called charging by friction even though friction actually has nothing to do with the charging process. Some materials tend to hold on to their electrons more strongly than others. When two materials that differ in this tendency are placed in intimate contact (i.e., rubbed together), one of them gains electrons and the other loses electrons.

The table below lists some materials in the triboelectric series (note: *tribo-* means friction), which ranks materials according to their tendency to give up their electrons. Materials toward the top of the list become positively charged when placed in contact with those lower on the list.

Table 13.1

	Positive
The Triboelectric Series	rabbit's fur
	glass
	wool
	cat's fur
	silk
	felt
	cotton
	wood
	cork
	rubber
	celluloid
	Negative

(a) If rabbit's fur is used to rub a piece of rubber, how do their charges change?

_____

(b) Would you rub felt on glass or on rubber to obtain a positive charge on the felt? _____

Answers: (a) The rubber turns negative and the fur turns positive.   (b) rubber (The glass would make the felt negative.)

12   INSULATORS AND CONDUCTORS

One object can cause another object to become charged by coming into physical contact with that object, if electrons move from one object to the other. The exchange of electrons, however, occurs only at the point of contact. The electrons received by an object being charged negatively may or may not distribute over the entire surface of the object, depending upon how good a **conductor** of electricity the object is. Some materials make good conductors because their electrons, rather than being bound to individual atoms, are free to move from one atom to another. The atoms in a good conductor naturally "share" their electrons so that when surplus electrons are added to one part of the material, electrons throughout the material will redistribute so that the effect is that of the charge being spread throughout the object.

A similar thing happens when a conductor is charged positively. When electrons are removed from one portion of the conductor, the remaining electrons redistribute so that the entire object is again charged.

Materials in which the electrons are not so free to move from atom to atom are **insulators**. When a part of an insulator receives excess charge, the charge remains on that part rather than distributing over the entire object. Most metals are good conductors, and most nonmetals are poor conductors—good insulators—of electricity.

Identify whether each of the following describes a conductor or an insulator.

(a) Added electrons redistribute themselves over an object. _____

(b) Could be used for sending electrical charges from one place to another.

_____

(c) Any excess charge remains in position. _____

(d) An object made up of atoms or molecules that easily share electrons.

_____

Answers: (a) conductor;   (b) conductor;   (c) insulator;   (d) conductor

13   THE ELECTROSCOPE

The following figure illustrates an **electroscope,** a device that is used to detect an electric charge. The heart of the electroscope is the metal strip down the center and the flexible metal foil whose top is attached to it. A cylindrical cover surrounds the metal foil and has glass windows across its ends to reduce the effect of air currents. The metal strip and foil are electrically insulated from the cylindrical cover. The cover serves only to keep out air currents that would disturb the delicate foil. Let's

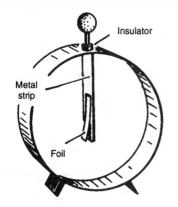

use our knowledge of charge to find out what happens when you touch a charged object to the ball at the top. The ball, metal strip, and foil are good conductors.

(a) What will happen to any charges that we transfer to the ball? _____

(b) Will the charges on the strip and foil be the same or different? _____

(c) What will happen to the foil? (Hint: It is very light.)

_____

Answers: (a) They will distribute over the ball, strip, and foil. (b) the same (both positive or both negative); (c) It will be repelled and will stand out from the metal strip.

14 As you see, the electroscope can detect the electric charge on an object that touches the ball. It also, however, removes that charge from the very object it is being used to examine. There is a way around this problem.

Suppose a negatively charged object is brought near—but not touching—the ball at the top of an uncharged electroscope.

(a) What type of force will electrons in the ball feel, attraction or repulsion?

_____

(b) What will these electrons try to do? _____

(c) Where can they go? _____

(d) What will be the resulting effect on the foil? _____

_____

(e) If you used a positively charged object near the ball, what type charge would get on the foil? _____

Answers: (a) repulsion; (b) leave, move away; (c) down the strip; (d) It will become negative and move away from the metal strip. (e) positive (as electrons move up the strip)

15 **CHARGING BY INDUCTION**

We discussed earlier charging by contact with a charged object. Now we will consider charging by **induction**. In this case an object is charged opposite to the object that charges it. Suppose that, while a negatively charged object is being held near the top of an electroscope, someone touches the ball. In this case some of those electrons that are trying to escape the vicinity of the charged object will be able to flow into the hand and down into the person's body. (The body is a fairly good conductor.) Now the hand is removed. This leaves a net positive charge on the electroscope. Notice that by using a negatively charged object here we put positive charge on an electroscope, causing some of the electrons to move onto another conductor that is in contact with the electroscope.

Can you describe how a positively charged object can be made to cause a negative charge on an electroscope—or any other conductor? _____

_____

_____

_____

Answer: The positively charged object is brought near the electroscope. Then when a person touches the ball, the electrons are pulled from the person's body, leaving excess electrons on the ball when the hand is removed.

SELF-TEST

The questions below will test your understanding of this chapter. Use a separate sheet of paper for your diagrams or calculations. Compare your answers with the answers provided following the test.

1. Suppose two charges are located 10 cm apart and feel an attraction force of 12 units. What will be the force between the two when they are placed 5 cm apart? _____

2. What is the SI unit of electric charge? _____

3. What do the arrows represent in a drawing of an electric field? _____

   _____

4. What is the direction of the electric field near a negative charge? _____

   _____

5. What is shown by the relative spacing of the lines in different parts of an electric field drawing? _____

   _____

6. Suppose the electric field at a particular location is 230 N/C. How much force will be exerted on a 0.03 coulomb-charge at that point? _____

7. What change takes place to make one object positive and the other negative when two objects are rubbed together? _____

   _____

8. Is it possible to make one object negative without making another object less negative? Why or why not? _____

    _____

9. What atomic particle moves when electricity moves through a conductor?

    _____

10. What happens to the electrons in a conductor after one part of the conductor is touched by a positively charged object? _____

    _____

    _____

11. How can an electroscope be charged negatively by induction? _____

    _____

    _____

12. *Optional.* Suppose two charges, each 4 microcoulombs, are located 20 cm apart. How much force is exerted on each charge? _____

13. *Optional.* How much charge must be on one object for it to be repelled with a force of 5 newtons by a second charge of 1 microcoulomb at a distance of 1 meter? _____

14. *Optional.* How much is the strength of the electric field at a distance of 20 cm from a 4-microcoulomb charge? _____

15. *Optional.* What will be the force on another 4-microcoulomb charge at the location for which you calculated the electric field in question 14? _____ Compare this answer to that of problem 12. _____

    _____

    _____

ANSWERS

If your answers do not agree with those given below, review the frames indicated in parentheses before you go on to the next chapter.

1. 48 units of force (The distance between them is $\frac{1}{2}$ as much. The inverse square of $\frac{1}{2}$ is 4, so the force becomes four times as great.) (frame 2)

2. coulomb (frame 3)

3. They show the direction of the force on a positive charge at that point. (frame 6)

4. It points radially inward toward the charge. (frame 7)

5. Where the lines are closer together, the field is stronger. There would be more force on a charge there. (frame 8)

6. 6.9 newtons. Solution: Multiply the electric field strength (230 N/C) by the charge located in the field (0.03 C). (frame 9)

7. Electrons leave the object that becomes positive and go onto the object that becomes negative. (frame 11)

8. No. An object is made negative by adding electrons to it. The electrons must be taken from some other object, which then becomes less negative (or more positive). (frame 11)

9. electrons (Protons are bound in place in the nuclei of atoms.) (frame 12)

10. Some electrons from the part of the conductor that is touched will move onto the positive object. Then electrons throughout the conductor will redistribute so that the lack of negative charge on the conductor is spread throughout. Thus, the positive charge is spread throughout. (frame 12)

11. A positively charged object is brought near the ball. Then the ball is touched by an uncharged conductor (such as your finger). This causes electrons to be pulled from the conductor onto the ball. Then the conductor is pulled away and the positive object is removed. The electroscope is negative. (frame 15)

12. 3.6 newtons. Solution: $\mathbf{F} = 9 \cdot 10^9 \cdot \dfrac{Q_1 \cdot Q_2}{d^2}$

$$\mathbf{F} = 9 \cdot 10^9 \cdot \frac{(4 \cdot 10^{-6})(4 \cdot 10^{-6})}{(0.2)^2}$$
$$= 3600 \cdot 10^{-3} \text{ newtons} \quad \text{(frames 4, 5)}$$

13. $5.56 \cdot 10^{-4}$ coulombs or 555 microcoulombs. Solution: Letting Q_2 be the unknown charge:

$$5 = 9 \cdot 10^9 \cdot \frac{(10^{-6}) \cdot Q_2}{(1)^2}$$
$$5 = 9 \cdot 10^3 \cdot Q_2$$
$$Q_2 = \frac{5}{9 \cdot 10^3}$$
$$= 0.556 \cdot 10^{-3} \text{ coulombs} \quad \text{(frame 5)}$$

14. $9 \cdot 10^5$ N/C. Solution: Letting Q_2 be the 1-coulomb charge that is put into the field:

$$\mathbf{F} = 9 \cdot 10^9 \cdot \frac{(4 \cdot 10^{-6})(1)}{(0.2)^2}$$
$$= 900 \cdot 10^3 \text{ newtons}$$
$$= 9 \cdot 10^5 \text{ newtons on 1 coulomb} \quad \text{(frame 10)}$$

15. 3.6 newtons. Solution: Multiply the electric field by the charge. This is the same answer that we obtained in problem 12, where we directly calculated the force on a 4 microcoulomb charge at a distance of 20 cm from another charge of 4 microcoulombs. (frame 10)

14 Electrical Current

Prerequisites: Chapter 13; Chapter 3, frames 13–17 (for frames 6, 19, and 20)
Electrons carry the charge in a charged object. When electrons move around in a conductor, we say that there is a current in the conductor. In this chapter we will see in more detail what happens when current flows. Most of the examples will use a metal wire as a conductor.

OBJECTIVES

After completing this chapter, you will be able to

- specify the direction of electron flow from a charged object to ground;
- differentiate between electron flow and conventional current flow;
- calculate the energy released when a given amount of charge flows across a given potential difference;
- specify the unit of current and relate it to charge and time;
- recognize schematics for batteries and resistors;
- use Ohm's law to solve simple circuit problems;
- determine total voltage across resistors connected in series and in parallel;
- determine total resistance when resistors are connected in series and in parallel;
- specify the effect of a burn-out (open circuit) in a series or parallel circuit;
- differentiate between alternating and direct current;
- determine the energy dissipated by a circuit, given the power and time;
- given two of the three quantities—current, voltage, and resistance—calculate the power dissipated in a circuit;
- given the cost of electrical energy, calculate the cost of operating a given device for a certain amount of time;
- calculate the number of joules in a kilowatt-hour (optional).

1 **FLOW OF ELECTRIC CHARGE**

Suppose that one end of a wire is connected to a very large conductor, one large enough so that we can add electrons to it or take them away without appreciably changing the density of charge on the object. (This can be done when the object is large enough that when electrons are added to it, they distribute themselves over it so that relatively few additional electrons are left in any one area.) Such an object is usually called "ground"; it may actually be the earth (which is, naturally, the best "ground" available). With one end of a wire connected to ground, suppose we touch a negatively charged object to the other end of the wire. When we do so, electrons move down the wire from the charged object to ground, creating an electric current. This current is very short in duration, lasting but a fraction of a second. Later we will consider methods by which a source of electrons is able to replenish its supply of electrons as they are removed, to produce a steady current in the wire. Examine the diagram below.

(a) What happens to electrons when the object touches the wire? _____

_____

(b) Why is the earth the "best ground available"?

_____

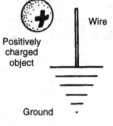

Answers: (a) They move from ground, up the wire, and into the object to balance the positive charge there. (b) It is the largest object around!

2

Physicists speak of the direction of "conventional" current in a conductor as the direction of the flow of *positive* charge. In other words, they say that electric current flows from the positive end of a wire to the negative end. This may create a momentary confusion. No one is claiming that protons move through the wire, however. We know that the electrons are the moving objects in a current-carrying wire.

Because it is more convenient to consider current flowing from positive to negative, and because the names "negative" and "positive" were assigned before it was known that electrons are the current carriers, we still speak in physics of current moving from positive to negative.

Refer back to the figure in frame 1.

(a) Which way do electrons move? _____

(b) Which way does conventional current flow? _____

Answers: (a) up the wire (negative to positive); (b) down the wire (positive to negative)

3 In the last chapter we saw that the unit of electric charge is the coulomb. The unit of electric current is the ampere. One ampere flows past a point in a conductor when 1 coulomb of charge moves by that point each second. Thus,

$$1 \text{ ampere} = 1 \text{ coulomb/second}$$

(a) How many amperes of current are caused by 2 coulombs of charge flowing in 0.1 second? _____

(b) How many amperes result from 60 microcoulombs flowing in 2 seconds? (1 microcoulomb = 10^{-6} coulombs) _____

Answers: (a) 20; (b) $3 \cdot 10^{-5}$ amp (0.00003 amp or 30 microamps) (See Appendix I for a review of powers of ten.)

4 BATTERIES AND VOLTAGE

Batteries are a familiar source of electric charge that causes current to flow in wires. Chemical reactions that take place in the batteries separate negative and positive charges, causing one terminal of the battery to be positive and the other negative. When the two terminals of the battery are connected (by a wire), electrons flow. If this flow is not too great, chemical action in the battery continues to take electrons from one terminal and place them on the other, maintaining the original charge.

(a) From which terminal are electrons taken by the chemical action of the battery? _____

(b) From which terminal do electrons leave to flow through the wire? _____

_____

(c) In what direction does the conventional current flow through the wire?

_____

Answers: (a) positive; (b) negative; (c) positive to negative

5 You may have noticed that batteries are marked in "volts." The common flashlight battery, for example, is 1.5 volts. The voltage of a battery tells us how much more energy each coulomb of charge has at one terminal than it has at the other. In moving from one terminal to the other, the charge loses this energy, which may be converted to heat and light. Thus, when 1 coulomb of positive charge flows from the positive to the negative terminal of a 1.5-volt battery, it gives up a certain amount of energy. If 2 coulombs flow across the 1.5-volt battery, twice as much energy is lost. Three coulombs—three times the energy given up. The voltage is equal to the energy given up by each coulomb of charge; it is the energy per coulomb.

(a) If 1 coulomb of charge flows across a 3-volt battery, how does the energy given up compare to that given up by 1 coulomb moving across a 1.5-volt battery? _____

(b) Does an electron have more energy at the positive or negative terminal?

_____

(c) Does positive charge have more energy at the positive or negative terminal?

_____

Answers: (a) twice as much; (b) negative (since it gives up energy as it flows to the positive terminal); (c) positive

6

The term *voltage* may describe energy per charge, but the more technical term is **potential difference**. The potential difference across the terminals of the standard flashlight battery is 1.5-volts.

The amount of energy a coulomb of charge loses when it falls across a potential difference of 1 volt is 1 joule. (Positive current naturally goes from positive to negative, like an object naturally falls from higher to lower elevations. By analogy, we often speak of electric charge "falling.")

$$1 \text{ volt} = 1 \text{ joule/coulomb}$$

(a) Three coulombs of charge flow across a 1.5-volt battery. How much energy is lost by the electric charge? _____

(b) How much charge must flow between the terminals of a 3-volt battery to release 12 joules of energy? _____

Answers: (a) 4.5 joules (3·1.5); (b) 4 coulombs

7 OHM'S LAW

Normally, one does not connect a wire straight from one terminal of a battery to the other, because this would cause too much current to flow, and the battery could not restore the charge to its terminals at the rate necessary. The battery would "run down." Besides, the circuit would accomplish nothing useful. Suppose, instead, that a lightbulb is placed in the circuit as shown in part 1 of this figure.

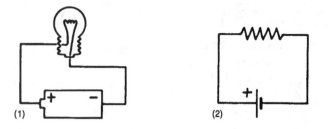

(1)   (2)

We usually use symbols to represent the various circuit components. Part 2 shows the circuit schematically. The saw-toothed line is used to represent the light-bulb in this case, but it may represent any device that offers resistance to the flow of electric current and limits the amount of current that will flow when a certain voltage is applied. We call this component a **resistor**.

What effect would you expect a resistor of greater resistance to have on current flow? _____

Answer: Cut it down (reduce it).

8

The relationship between voltage, current, and resistance is given by Ohm's law:

voltage (in volts) = current (in amperes)·resistance (in ohms)

or, symbolically:

$$V = I \cdot R$$

where I represents electric current. Resistance (R) is measured in ohms (pronounced to rhyme with "homes" and symbolized by Ω).

Although potential difference (voltage) is defined as being the energy per coulomb, it can be thought of loosely as being the "pressure" that causes the electric current to flow. Just as more water pressure causes an increase in the amount of water that flows through a pipe, more voltage causes more current to flow. But when resistance is increased, less current flows.

(a) Express the unit "ohm" in terms of amps and volts. _____

(b) Suppose you use a 3-volt battery to light a lamp that has a resistance of 30 ohms (symbolically: 30 Ω). Use $V = I \cdot R$ to determine how much current flows.

_____

Answers: (a) volts/amp;   (b) 0.1 ampere (or amp). Solution: $I = V/R$
$$= 3 \text{ volts}/30 \ \Omega$$

9 BATTERIES IN SERIES AND PARALLEL

Most flashlight bulbs are designed so that when 3 volts are connected across them they burn at their designed brightness. (If they burned hotter, they would burn out too quickly, and if they burned cooler, they would not provide as much light.) Since the common flashlight battery is 1.5 volts, we connect two batteries as shown in part 1 of the figure here in order to light the 3-volt flashlight bulb. Part 2 shows this in schematic form. (Notice the symbol for a battery.) The positive terminal of battery X is 1.5 volts higher in voltage than the negative terminal, and the negative terminal of battery Y is connected directly to this. Then the positive terminal of this battery is another 1.5 volts higher in voltage. Batteries connected positive-to-negative like this are said to be connected "in series."

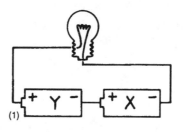

(1)

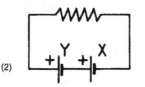

(2)

(a) What is the total voltage (potential difference) across these two batteries in series? _____

(b) In general, how would you find the overall voltage of batteries connected in series? _____

Answers: (a) 3 V; (b) Add the individual voltages.

10 The figure in this frame shows two batteries hooked up side-by-side (positive-to-positive and negative-to-negative). Batteries connected like this are said to be "in parallel." The potential difference across the resistor (which may be a lightbulb) when batteries are connected like this is the same as the potential difference across one of the batteries. The advantage of such a hookup is that the batteries will last longer. That is, they will be able to push more charge through the light-bulb before running down.

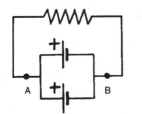

(a) How does the effective voltage of two batteries connected in series compare to that of the same two batteries connected in parallel? _____

(b) If the batteries of the figure are each 1.5 volts, what is the potential difference between points A and B?

_____

Answers: (a) double;   (b) 1.5 volts

11 RESISTORS IN SERIES AND PARALLEL

The figure here shows three resistors connected in series. In a series connection, current must go through all three of the resistors to get from one side of the battery to the other. The total resistance of a series circuit is the sum total of the resistances of each of the resistors.

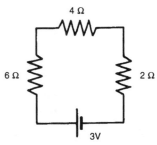

(a) What is the total resistance of the circuit of the figure in this frame? _____

(b) Use Ohm's law to find how much current will flow. _____

Answers: (a) 12 ohms;   (b) 0.25 amp ($I = V/R = 3/12 = 0.25$)

12 The total voltage across a number of resistors in series is equal to the sum of the voltage drops across the individual resistors. Let's calculate the potential difference across just one of the resistors of the figure in the previous frame. Consider the 2-ohm resistor. The current in it is 0.25 amp. (All of the current flows through the entire circuit.) The voltage necessary to force 0.25 amp through 2 ohms is found using Ohm's law again:

$$V = 0.25 \text{ amp} \cdot 2 \text{ ohms}$$
$$= 0.50 \text{ volt}$$

(a) What is the voltage across the 4-ohm resistor? _____

(b) What is the voltage across the 6-ohm resistor? _____

(c) What is the sum of the voltages across all three resistors? _____

(d) How does this compare to the voltage across the battery? _____

Answers: (a) 1 volt;   (b) 1.5 volts;   (c) 3 volts;   (d) the same

13 The figure here shows resistors connected side-by-side, "in parallel." In a parallel circuit, current leaving one terminal of the battery has a choice of paths that it can take to get to the other terminal. In this case, each resistor has a resistance of 6 ohms. Since the current has three paths to take, it divides among those three. The effective resistance of resistors in parallel is always less than the resistance of any one of the resistors. The effective resistance R in a parallel circuit is given by the equation:

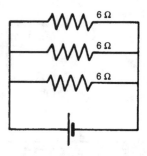

$$\frac{1}{R} = \frac{1}{R_1} + \frac{1}{R_2} + \frac{1}{R_3} + \ldots$$

where R_1, R_2, and so on are the resistances of the individual resistors.

Calculate the effective resistance of the three resistors in the figure in this frame. _____

Answer: 2 ohms. Solution: $\frac{1}{R} = \frac{1}{6} + \frac{1}{6} + \frac{1}{6} = \frac{3}{6}$; $R = \frac{6}{3} = 2$

14 Suppose you have two resistors in parallel, one 2 ohms and the other 3 ohms.

(a) Will the effective resistance be less than or greater than 2 ohms?

_____

(b) What is the value of the effective resistance? _____

Answers: (a) less than;   (b) $\frac{6}{5}$ ohms

15 Follow the steps below to solve the problem represented by this circuit.

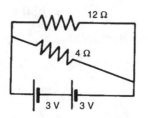

(a) Are the resistors in series or in parallel? _____

(b) Are the batteries in series or in parallel? _____

(c) How much is the potential difference across the battery combination? _____

(d) How much is the resistance of the circuit? _____

(e) How much current flows from the batteries? _____

(f) How much current flows through the 12-ohm resistor? _____

(g) How much current flows through the 4-ohm resistor? _____

(h) Is your answer to question (e) equal to the total of your answers to (f) and (g)? _____   Why? _____

Answers: (a) parallel*;   (b) series;   (c) 6 volts;   (d) 3 ohms;   (e) 2 amps (6 volts/3 ohms); (f) 0.5 amp (6 volts/12 ohms);   (g) 1.5 amps;   (h) Yes. The current divides to go through the two resistors and the sum of the parts equals the whole.

16   SERIES VS. PARALLEL: HOME WIRING

Let's consider what the effect on a circuit is when a lightbulb burns out. A bulb burns out when the tiny hot wire—the filament—inside it breaks. (The molecules of the filament gradually evaporate directly from solid to gas as the lamp burns. Thus, the filament gets thinner and thinner until it breaks.) Now look at the figure back in frame 11, the circuit with three lightbulbs in series. Imagine the wire within one of the lamps (the resistors) breaking.

(a) What happens to the other two lamps? _____

(b) Does it make any difference which lightbulb burns out? _____

(c) Suppose all your Christmas tree lights go out when a single bulb is removed. Are these wired in series or in parallel? _____

Answers: (a) They go out.   (b) no (They all go off no matter which burns out.);   (c) series

17   Now consider the parallel circuit of the figure in frame 13. When one of the bulbs burns out in this circuit, current can still flow through the other branches. The fact that current no longer flows through one branch of a parallel circuit does not

---

*Notice that "in parallel" does not mean that the resistors are geometrically parallel to one another, but rather that electrical charge has different paths to get from the positive side to the negative side of the battery. (In this sense, the interstate highway I-75 runs "parallel" to I-65 to carry traffic from north to south.)

prevent it from getting from one terminal of the battery to the other by means of the other circuits.

(a) If two bulbs burned out in this circuit, what would happen to the third?

_____

(b) Are the circuits in your home wired in series or in parallel? _____

Answers: (a) It would continue to burn.   (b) parallel (When one electrical device burns out—or is unplugged or turned off—the remainder still work.)

18   ALTERNATING CURRENT

We have been discussing direct current (DC), the kind supplied by a battery. The current we use every day in our homes is alternating current (AC). Both types of current consist of electrons moving through the wire, but in house wiring the electrons move first one direction, then the other. In order for this to happen the ends of the wires supplying the current alternately change back and forth between positive and negative.

The alternating current system used in the United States changes from positive to negative and back to positive 60 times each second. Its frequency is therefore 60 cycles per second, or 60 hertz. (*Hertz* is a single word that means "cycles/second.") In Europe, 50-cycle/second alternating current is used.

As far as lightbulbs are concerned, the difference between direct and alternating current is unimportant, for it is the flow of current that makes the filament heat up. It does not matter to the bulb whether or not the current changes direction. Devices that contain motors, or electronic devices such as radios, stereos, and TVs, must be designed to work either with AC power or DC power. If you have such a device that can be used with either power source, it contains a circuit that changes one type of current to the other. (Normally the AC is changed into DC when portable radios or CD players are plugged into a socket in your home.)

(a) What type of current is used in a portable calculator? _____

(b) What type of current is used to run an electric table saw? _____

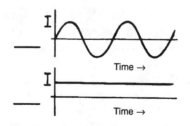

(c) Two graphs are shown in the figure, each plotting current as time passes. Label each of the graphs on the previous page as representing AC or DC.

Answers: (a) direct; (b) alternating; (c) top: AC, bottom: DC

19 ENERGY DISSIPATED IN A RESISTOR

We said earlier that voltage is the energy exchanged per coulomb of charge. As each coulomb flows through a resistor, this energy shows up in the resistor as heat (and light, if the resistor gets hot enough—as in the lightbulb). The rate at which energy is transferred to heat and light is called power, and the unit of power is the watt.

$$\text{power} = \text{energy/time}$$

A 40-watt lightbulb will use less energy than a 60-watt bulb in the same amount of time. To calculate the amount of energy used, we multiply power by the time during which the power is used. If power is expressed in kilowatts (1 kilowatt = 1000 watts) and time in hours, then the unit of energy is kilowatt-hour (abbreviated kWh).

(a) How many kilowatt-hours are used in burning a 100-watt bulb for 8 hours?

_____

(b) Since the unit of kilowatt-hours appears on an electric bill, are you paying for power or energy? _____

(c) Suppose you pay 15 cents per kilowatt-hour. How much does it cost to watch a 300-watt TV for 6 hours? _____

Answers: (a) 0.8; (b) energy; (c) 27 cents. Solution: 300 W = 0.3 kW; 0.3 kW·6 hr = 1.8 kWh; 1.8 kWh·15 cents/kWh = 27 cents

20 SI CALCULATIONS

As you recall, 1 volt = 1 joule/coulomb and 1 ampere = 1 coulomb/second. When we multiply potential difference (volts) by current (amperes), we obtain:

$$\frac{\text{joules}}{\text{coulomb}} \cdot \frac{\text{coulomb}}{\text{second}} = \text{joules/second}$$

One joule/second = 1 watt. The product of the potential difference across a resistor and the current through the resistor is the power input into the resistor. In equation form:

$$P = I \cdot V$$

Ohm's law is $V = IR$. Thus, when we substitute this into the power equation we obtain $P = I^2R$.

From Ohm's law, $I = V/R$. Substituting V/R for I in the first equation, we obtain $P = V^2/R$.

These three equations for power are equivalent, but sometimes one is more convenient to use than another.

(a) To calculate the resistance of a lightbulb that, when used in a 115-volt circuit, uses 100 watts of power, which formula will you use? _____

(b) What is the numerical answer to this problem? _____

(c) How much current flows through the lamp? _____

Answers: (a) $P = V^2/R$;   (b) 132 ohms;   (c) 0.87 amps (From $P = IV$ or $V = IR$ or $P = I^2R$. The first is preferred because it does not rely on your calculation of R.)

21   *Optional.* Work through the steps below to discover the number of joules in a kilowatt-hour.

(a) How many watt-hours are in a kilowatt-hour? _____

(b) How many seconds are in an hour? _____

(c) If a watt is a joule/second, what is a joule (in terms of watts and seconds)?

_____

(d) One kilowatt-hour is how many joules? _____

Answers: (a) 1000;   (b) 3600;   (c) 1 watt-second;   (d) $3.6 \cdot 10^6$, or 3,600,000

SELF-TEST

1. One-half ampere flows from one terminal of a 3-volt battery to the other terminal for 5 seconds. How much energy is released?_____

2. What does the voltage of a battery tell us about the electric charge on the terminals of the battery? _____

_____

3. Suppose a 9-volt battery is connected across a 100-ohm resistor. How much current will flow through the resistor? _____

4. If a greater potential difference is applied across a resistor, _____ (more/the same/less) current will flow.

5. Are the batteries in a conventional flashlight connected in series or in parallel?

_____

6. A "power beam" flashlight contains five regular D-cell batteries (1.5 volts each) in series. What is the voltage across these five batteries?

_____

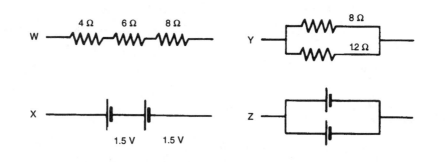

7. (a) Which of the parts in the figure represent resistors in series? _____

   (b) Which represents batteries in series? _____

   (c) Which represents resistors in parallel? _____

8. (a) How much is the effective resistance of part Y of the figure? _____

   (b) How much is the effective resistance of part W of the figure?_____

9. How does the effective resistance of four resistors in parallel compare to the individual resistances? _____

   _____

10. As more and more resistors are added in parallel, what happens to the total current flowing from a source? (This occurs when you continue to plug in more and more devices in your home.) _____

    _____

11. If electrical energy costs 10 cents per kilowatt-hour, how much does it cost to operate a 1200-watt hair dryer for a half hour?_____

12. Suppose an auto headlight is on for 1 hour. Two amperes of current flows through it from the 12-volt battery of the car.

    (a) How many coulombs of charge flow through the headlight during the hour? _____

    (b) How much energy does this charge release (in joules)? _____

    (c) How much is the power in watts of the headlight used like this?_____

13. One-half amp flows through a 40-ohm resistor.

    (a) How much power is being used? _____

    (b) How much is the voltage across the resistor? _____

14. Refer to the figure in question 7. Suppose the battery connection of part X is connected to the resistors of part W.

    (a) How much current flows? _____

    (b) How much is the potential difference across the 6-ohm resistor?

    _____

ANSWERS

1. 7.5 joules ($\frac{1}{2}$ amp is $\frac{1}{2}$ coulomb each second. In 5 seconds this is 2.5 coulombs, and since the potential difference is 3 volts, or 3 joules/coulomb, this amounts to $3 \cdot 2.5$, or 7.5 joules.) (frames 3, 6)

2. It tells us how much more energy the charge has on one terminal than on the other. Voltage is energy per charge. (frame 5)

3. 0.09 ampere ($V = IR$; $9 = I \cdot 100$) (frame 8)

4. more (This follows from Ohm's law.) (frame 8)

5. series (frame 9)

6. 7.5 volts (frame 9)

7. (a) W (frame 11); (b) X (frame 9); (c) Y (frame 13)

8. (a) 4.8 ohms. Solution:

$$\frac{1}{R} = \frac{1}{R_1} + \frac{1}{R_2}$$
$$= \frac{1}{8} + \frac{1}{12}$$
$$= \frac{3}{24} + \frac{2}{24}$$
$$= \frac{5}{24}$$
$$R = \frac{24}{5} = 4.8 \qquad \text{(frames 13, 14)}$$

   (b) 18 ohms (frame 11)

9. The effective resistance is less than the resistance of any of the resistors. (frame 14)

10. The total current increases, because the effective resistance decreases. (In your home, too much current will blow a circuit breaker.) (frames 13–15))

11. 6 cents. Solution:
$$1200 \text{ watts} = 1.2 \text{ kilowatts}$$
$$1.2 \text{ kW} \cdot 0.5 \text{ hr} = 0.6 \text{ kWh}$$
$$0.6 \text{ kWh} \cdot 10 \text{ cents/kWh} = 6 \text{ cents (frame 19)}$$

12. (a) 7200 coulombs (2 amps = 2 coulombs/second; 1 hr = 3600 s) (frame 3)

(b) 86,400 joules (7200 coulombs · 12 joules/coulomb) (frame 6)

(c) 24 watts ($P = IV$) (frame 20)

13. (a) 10 watts ($P = I^2R = 0.5^2 \cdot 40 = 0.25 \cdot 40$)

(b) 20 volts ($V = IR = 0.5 \cdot 40 = 20$ volts) (frame 20)

14. (a) 0.167 amp ($V = IR$, $I = \dfrac{V}{R} = \dfrac{3}{18} = 0.167$ amp)

(b) 1 volt ($V = IR = 0.167 \cdot 6 = 1$ volt) (frames 8, 9, 11, 12)

15 Magnetism and Magnetic Effects of Currents

Prerequisite: Chapter 14
A magnet has a north and a south pole; like poles repel and unlike poles attract. This rule of attraction and repulsion is very similar to the one that applies to electric charges. (In fact, the equation for the force between two magnetic poles has the same form as the Coulomb's law equation—and as the law of gravity.) The connection between magnetism and electricity is much closer than just the similarity of force laws, however, and most of this chapter will be spent discussing this connection.

OBJECTIVES

After completing this chapter, you will be able to

- interpret a magnetic field drawing in terms of relative strength and directions of forces on north and south poles;

- describe the magnetic field of the earth and specify what types of magnetic poles are located at the earth's north and south geographic poles;

- use the dot-and-X convention to indicate directions of electric currents and magnetic fields;

- indicate the direction of a magnetic field produced when current passes through a wire;

- specify the direction of a magnetic field due to current in a loop or solenoid;

- state the effect of an iron core in a solenoid;

- state the difference between a magnetized and an unmagnetized object in terms of domains;

- describe the operating principles of voltmeters or ammeters;
- describe the operating principles of electric motors;
- state the nature of the force by which both electric meters and motors operate;
- specify, on the atomic level, what causes magnetism in magnetic materials;
- describe the method used to change current direction in the loop of an AC motor and a DC motor;
- calculate the strength of magnetic field in the case of long straight wires and in the case of solenoids (optional).

1   MAGNETIC FIELDS

The figure shows a bar magnet with its north and south poles marked. The lines with arrowheads represent the magnetic field of the magnet. We define the direction of this field at any point to be the direction of the force on a north pole placed at that point. As in the case of an electric field, the relative density of the lines tells us the relative strength of the field.

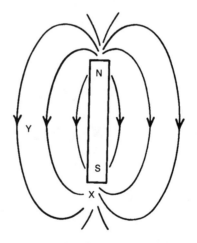

(a) At which point, X or Y, would another north pole experience the greater force? _____

(b) Would the force that is exerted on a north pole at point Y be upward or downward on the page? _____

Answers: (a) X (greater density of lines);   (b) downward

2

Suppose we have a small bar magnet supported from its center so that it is free to swing about that center. Now suppose we locate this magnet at point Y in the magnetic field of the figure in the previous frame. The north pole of this small suspended magnet will feel a force downward, along the direction of the arrow indicated on the lines there. The south pole will feel a force in the opposite direction. The suspended magnet will therefore align itself along the magnetic field of the bar magnet. In practice, this is one method used to plot the magnetic field of a magnet. (You might do such an exercise in an elementary physics laboratory.) A more common use of such a device is as a compass.

The figure of this frame shows that the earth has a magnetic field, in a form similar to that of a bar magnet. When a magnet pivoted at its center—a compass—is held in this field, the north pole of this magnet points in a direction that we call north. If you have ever used a magnetic compass, you know that the end of the compass that we call "north" points toward the geographic north of the earth.

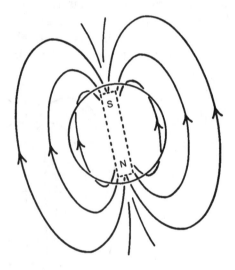

What type magnetic pole is located near the geographic north pole of the earth? _____

Answer: a south magnetic pole (This probably seems strange, but if we call the end of the compass that points north a north magnetic pole, no other answer is possible, for opposite poles attract.)*

---

*Some people resolve this conflict of terminology by saying that the end of the compass needle that points northward is the "north-seeking" pole of the compass, and that whenever someone refers to the north pole of a magnet (except for the north pole of the earth), he or she is actually talking about a north-*seeking* pole. In any case, the end of the compass that points northward has the opposite type magnetic pole of the one at the geographic north pole.

FIELDS PRODUCED BY CURRENTS

A bar magnet is only one way to produce a magnetic field. Magnetic fields are produced every time an electric current flows in a wire. Part 1 of the figure shows a wire with a current flowing upward. The circles represent the magnetic field that surrounds the wire. The field's direction is around the wire, perpendicular to the direction of the current. In front of the wire (between you and the page) the direction of the field is from left to right, while on the back side of the wire the field is from right to left.

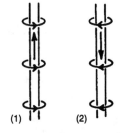

(1)   (2)

Part 2 of the figure shows a current going downward in a wire. How does the direction of the magnetic field in part 2 compare to that in part 1? _____

Answer: opposite direction

The important thing about the magnetic field near a current-carrying wire is that the field is circular, around the wire. All that is left, then, is to decide which way it points around the wire. You can determine the direction of the field caused by a current in a wire by imagining that you grab the wire with your right hand so that your thumb points in the direction of the current. Your fingers will then curve in the direction of the magnetic field. (We'll call this technique the "curled fingers right-hand rule.") Try this to verify the directions of the fields in the two parts of the figure in the previous frame.

Part 1 of the figure in this frame shows another method of representing magnetic fields. This figure shows a current flowing from left to right through the wire. Using the right-hand rule, you see that the field above the wire (at point A) is coming out of the page. Such a field is represented by dots. The field below the wire, pointing into the page, is represented by X's. This method is easily remembered by thinking of actual arrows (as shot from a bow) with feathers at their trailing end. As the arrow moves away from us, we see the feathers as an X, and when it is coming toward us we see its point.

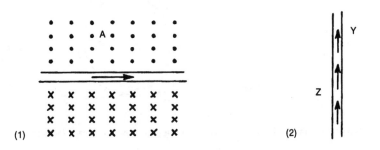

(1)   (2)

(a) Use the right-hand rule to determine the direction of the magnetic field at Y near the wire in part 2 of the figure._____

(b) What is the direction of the magnetic field in the air midway between you and the wire in this case?_____

(c) What would be the direction of the force on a north pole at point Z?

_____

Answers: (a) down into the page;   (b) left to right;   (c) out of the page, toward you

5   STRENGTH-OF-FIELD CALCULATIONS (OPTIONAL)

The unit of magnetic field is the tesla (T).* The equation for the magnetic field B (in tesla) at a distance r (in meters) from a wire carrying a current I (in amperes) is:

$$B = 2 \cdot 10^{-7} \cdot \frac{I}{r}$$

Thus, when 1 ampere flows in a long, straight wire, the magnetic field 1 meter from the wire is $2 \cdot 10^{-7}$ tesla. What is the magnetic field at a distance of 1 centimeter from a long, straight wire carrying a current of 8 amperes? _____

_____

Answer: $1.6 \cdot 10^{-4}$ T (Be sure to express distance in meters.)

6   MAGNETIC FIELDS DUE TO LOOPS AND SOLENOIDS

Imagine grabbing the wire of the loop shown in this figure with your right hand so that your thumb points along the current direction. You will find that no matter where you grab the wire of the loop, the magnetic field inside the loop points up out of the paper. This tells you that every part of the loop contributes to the field coming out of the center of the loop.

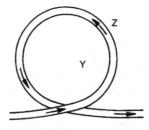

_____

*An older name for this unit is the weber per square meter (Wb/m²).

(a) What is the direction of the field at point Y? _____

(b) At point Z? _____

Answers: (a) out of the page toward you; (b) into the page (Although the contribution at Z due to the left side of the loop is opposite to this, it is weaker because the left side of the loop is farther away.)

7

The figure in this frame shows a solenoid, which can be thought of as a number of loops stretched out along the axis of the loops. Again, each portion of each loop contributes to the field through the center. Because of this the field inside a solenoid can be much more intense than the field near a long wire or a single loop of wire.

Notice that the magnetic field external to the solenoid is similar to that of the bar magnet of the figure in the first frame of this chapter. If a metal that is easily magnetized and demagnetized (soft iron, for example) is placed within a solenoid, the metal will become magnetized when current flows through the wire of the solenoid. The additional magnetism due to the metal will then contribute additional magnetic field to the region surrounding the solenoid. The solenoid and enclosed iron core thus forms a very strong magnet. Such **electromagnets** are used on the end of large cranes to pick up magnetic materials.

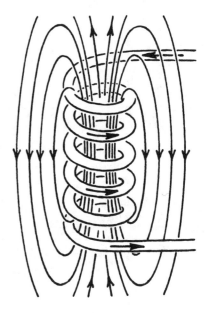

(a) Would a solenoid with no current flowing in its wire function as a magnet?

_____

(b) Would a solenoid without a core function as a magnet? _____

(c) How would you stop an electromagnet from acting as a magnet?_____

_____

Answers: (a) no (except for possible residual magnetism in the iron core due to current that previously existed); (b) Yes, but it would not be as strong. (c) turn off the current

8 CALCULATING THE MAGNETIC FIELD OF A SOLENOID (OPTIONAL)

The magnetic field at the center of a long solenoid of n turns and length l is given by the equation:

$$B = 4\pi \cdot 10^{-7} \cdot \frac{nI}{l}$$

The fact that the diameter of the solenoid is not included in the equation shows that the diameter does not really affect the field. (The diameter must be small compared to its length, however. This is why we specify a long solenoid.)

A typical 10-cm-long solenoid might have 1000 turns of wire carrying $\frac{1}{2}$ amp. What is the magnetic field in the center of such a solenoid?

Answer: $6.3 \cdot 10^{-3}$ T. Solution: $B = 12.6 \cdot 10^{-7} \cdot \dfrac{1000 \cdot 0.5}{0.1}$

9 THE CAUSE OF MAGNETISM

Let's consider what is different about a material when it is magnetized. Such magnetism is caused by an electric current moving in a circle, a loop. We saw that when current moves in a circle, there is a magnetic field through the loop, making the loop act like a small magnet. In an atom, the electric charges (electrons) move in a circle about the nucleus, and in addition, each electron spins on its own axis. The motion of the electron in a circular path results in a magnetic field, and the spinning of the electron forms an additional magnetic field. In atoms of most materials, the magnetic fields produced by these revolving and spinning electrons cancel one another out by revolving and spinning in opposite directions, but a material that is magnetic has atoms with unbalanced magnetic fields. Thus, the atom itself acts as if it were a magnet. The magnet-atoms tend to align themselves so as to form small magnetic areas within a material. These areas are called **domains**, or **magnetic domains**. When the north and south poles of the various domains are arranged randomly with respect to one another, the piece of metal is not magnetized. This is shown in figure A on the next page. When the piece of magnetic material is placed in a strong magnetic field—perhaps caused by another magnet—these domains swing around so that they align with the north pole of one contacting the south pole of another. Figure B shows a perfectly ordered arrangement that causes the material to act as a magnet.

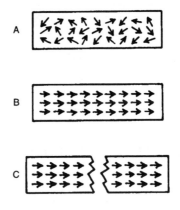

(a) Suppose you break a permanent bar magnet into two pieces, as in figure C. Is each piece a separate pole, so that one piece is north and the other piece south? _____

(b) Suppose figure A were inside a solenoid. What would happen to its domains when current was turned on? _____

Answers: (a) No, each piece has a north and a south pole. (Notice that on the broken end of the left fragment, there are exposed north poles of domains, and south poles show on the broken edge of the right fragment.) (b) They would line up (as shown in figure B).

10 ELECTRIC METERS

Consider what would happen if a current was passed through the loop of the figure at the top of the next page. The loop is between north and south magnetic poles (which could either belong to a horseshoe magnet or be parts of two separate magnets). When current flows as shown, a magnetic field directed out of the page will be formed by the current. So we can consider the current-loop to act like a small magnet with its north pole above the plane of the paper. (Recall from frame 7 the similarity between the field of a loop or solenoid and that of a bar magnet.) Such a magnet would experience a force on its north pole toward the south pole of the surrounding magnet and a force on its south pole toward the north pole of the surrounding magnet. Thus, the loop feels a twist on it, and if it is free to do so, it will turn 90° and align with the larger magnet. If, instead of one loop, there are a number of loops of wire between the poles of the permanent magnet, the loops will feel an even greater twist.

In a meter designed to detect or measure electrical current, however, the loop is restrained by a spring from turning all the way into alignment. A pointer is connected to the loop so as to indicate the amount of twist produced by the current through the wire of the loop. Both ammeters (to measure current) and voltmeters

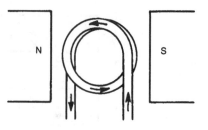

(which measure voltage by the amount of current that flows through them) work on this principle.

(a) Consider the loop of wire in an electrical meter. What type of force stops it from twisting to align with the poles of the permanent magnet? _____

_____

(b) What determines the amount of magnetic force that twists the loop?_____

_____

(c) How does increased current affect the twisting force?_____

Answers: (a) force exerted by the spring;   (b) the strength of the large magnet and the current in the loop (and the number of turns of wire in the loop);   (c) increases it

11 ELECTRIC MOTORS

It is a fairly small step from a meter to a motor. In the motor there is no spring to keep the loop (or coil) of wire from rotating, so it rotates until it aligns with the large magnet. Its connection to a power supply is such that, when it achieves this alignment, the direction of the current through the loop is reversed. This causes the loop to be forced away from that position and around 180° farther. At this point the current direction is again made to change so that the loop must again rotate 180° to come back into alignment.

Two methods are used to achieve the change in direction of current. The one used on motors designed for use with DC current is shown in the figure. Each end

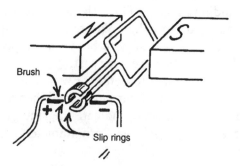

of the wire loop is connected to a semicircular piece of metal (called a "slip ring") into which the current flows from contacts (called "brushes") that rub it. The current enters the ring that is in contact with the left brush (marked "+"). It then passes through the loop and emerges from the other ring into the brush touching it. Since the rings turn with the loop, first one slip ring and then the other is in contact with the left brush. Thus, the current through the loop changes direction every time the loop turns through 180°.

The current can be made to change more easily by using alternating current, which regularly changes direction. Since the speed of rotation of the loop is dependent upon the frequency of alternation of the current, such motors are used where a regulated speed is important, as in clocks, for example. AC motors that are intended to produce power rather than regulated speed are wired differently, using electromagnets controlled by the current rather than permanent magnets around the coil.

(a) What is the basic difference between a meter and a motor? _____

_____

(b) Is it true that a motor run by DC must have current changing direction regularly through the loop? Why, or why not? _____

_____

Answers: (a) In the motor, the loop spins freely and current direction changes.   (b) yes; in order to have a changing magnetic field produced by the loop

SELF-TEST

1. What is the general law of attraction and repulsion between magnetic poles?

_____

2. What defines the direction of the magnetic field?_____

_____

3. What does the relative density of lines of a magnetic field tell us about the strength of the field? _____

_____

4. What type magnetic pole is found near the geographic south pole of the earth? _____

5. Suppose current is flowing from the bottom of this page to the top along a wire running up the right side of the page. What is the direction of the field at the location of this question mark? _____

6. The figure below represents a loop of wire carrying current in a clockwise direction. What is the direction of the magnetic field in the center of the loop?

_____

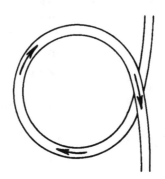

7. Refer back to the figure in frame 7. If this solenoid were to be replaced by a bar magnet to cause a similar field, would the north pole of the magnet be at the top or bottom? _____

8. What two actions do electrons have that cause materials to be magnetic?

_____

_____

9. What determines how far the loop of wire in an electric meter rotates when current is sent through it? What prohibits it from rotating into complete alignment with the external field? _____

_____

10. What are two ways in which the current in the coils of electric motors is made to change directions (in different types of motors)? _____

_____

_____

11. *Optional.* What current must flow in a long, straight wire in order to produce a magnetic field of 10^{-5} tesla at a distance of 2 centimeters from the wire?

_____

12. *Optional.* A 20-cm-long solenoid having 150 turns has a 2-amp current in it. What is the strength of the magnetic field at its center? _____

_____

ANSWERS

1. Like poles repel and unlike poles attract. (chapter introduction and frame 1)

2. The direction of the magnetic field at any point is the direction of the force on a north pole at that point. (frame 1)

3. Where the lines are more closely spaced, the field is stronger. (frame 1)

4. north (frame 2)

5. out of the page toward the reader (frame 4)

6. into the page, away from you (frame 6)

7. the top (frames 1, 7)

8. The electron spins on its own axis and each electron revolves around the nucleus of the atom. (frame 9)

9. the amount of current flowing (and the strength of the external magnet); the spring limits the rotation (frame 10)

10. Direct current can be made to change direction in the loop by slip rings rotating with the loop, or the motor can be designed to use the natural changes of current direction that exist in alternating current. (frame 11)

11. 1 amp. Solution: $B = 2 \cdot 10^{-7} \cdot \dfrac{I}{r}$

$$I = B \cdot \frac{r}{(2 \cdot 10^{-7})}$$

$$I = 10^{-5} \cdot \frac{0.02}{(2 \cdot 10^{-7})} \qquad \text{(frame 5)}$$

12. $1.89 \cdot 10^{-3}$ tesla. Solution: $B = 12.6 \cdot 10^{-7} \cdot \dfrac{nI}{l}$

$$= 12.6 \cdot 10^{-7} \cdot 150 \cdot \frac{2}{0.2} \qquad \text{(frame 8)}$$

16 Electrical Induction

Prerequisites: Chapter 14 and 15
In the last chapter we saw that when an electrical current exists in a wire, a magnetic field surrounds the wire. In this chapter we will see that a simple expansion of this idea allows us to explain the phenomenon of currents being produced by magnetic fields.

OBJECTIVES

After completing this chapter, you will be able to

- use the right-hand rule to determine the direction of force on a current-carrying wire in a magnetic field;

- determine the direction of induced current in a wire moving in a magnetic field;

- specify the effect a changing magnetic field has on a conductor;

- use Lenz's law to determine the direction of induced current;

- describe what happens when direct current is turned on and off through one of the coils of a transformer;

- specify the reason an iron core is used in a transformer;

- relate the number of turns in the primary and secondary of a transformer to the voltage across each, and given three of the above quantities, calculate the fourth;

- apply the principle of conservation of energy to a transformer, showing that power cannot be increased;

- identify the effect of self-induction in a coil;

- distinguish between the effect of an inductor on a high-frequency AC and a low-frequency AC.

1 A CURRENT-CARRYING WIRE IN A MAGNETIC FIELD

In the previous chapter (frame 11), I described how a motor worked by considering the coil of a motor to act like a small magnet. That magnet, I said, is pulled around by a large permanent magnet that surrounds it. A simple example of the same phenomenon is shown in the figure of this frame. Here a wire passes between the poles of a horseshoe magnet. A current is passing through the wire and is directed away from the reader. If you try this, you will find that there is an upward force on the wire, tending to lift it out of the magnet. (The effect is fairly weak unless you have a strong magnet and a current of at least a few amps.)

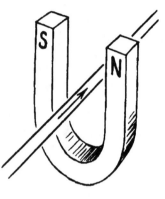

(a) In what direction would the force be if the current were reversed?

_____

(b) What would happen if you reversed the current and also reversed the magnet?

_____

Answers: (a) downward (into the magnet);   (b) The force would be upward (just as it was before the change).

You might want to return to frame 11 of Chapter 15 and convince yourself that instead of considering the coil as a magnet, you get the same results by considering it as individual current-carrying wires in a magnetic field. The wire on one side of the coil feels a force upward and the wire on the other side is pushed downward.

2   The directions involved in the example of frame 1 can best be remembered by use of a memory device that uses your right hand again. Hold the thumb and first finger of your right hand so that they are mutually perpendicular, as shown in the figure in this frame. (When your thumb points up and the middle finger points to your chest, the index finger will point to your left—try it.) Now let the Middle finger represent the Magnetic field, the Index finger represent the current (symbolized by an *I*), and the THumb represent the direction of the force, or THrust. Use this rule to verify the direction of the force on the wire in the figure in the first frame. The wires of the coil on a motor are acted on by this same force and can be analyzed in like manner.

Now refer to the figure here that shows a magnetic field represented by X's. (The field might be caused by a large magnet that is not shown.)

$$
\begin{array}{ccccccc}
\times & \times & \times & \times & \times & \times & \times \\
\times & \times & \times & \times & \times & \times & \times \\
\times & \times & \times & \times & \times & \times & \times \\
\times & \times & \times & \times & \times & \times & \times \\
\end{array}
$$

$$\longleftarrow$$

$$
\begin{array}{ccccccc}
\times & \times & \times & \times & \times & \times & \times \\
\times & \times & \times & \times & \times & \times & \times \\
\times & \times & \times & \times & \times & \times & \times \\
\times & \times & \times & \times & \times & \times & \times \\
\end{array}
$$

(a) In what direction is the magnetic field in relation to the page? _____

_____

(b) Use the right-hand rule. What is the direction of the force on the wire?

_____

Answers: (a) into the page (Chapter 15, frame 4);   (b) down the page

3 ┃ INDUCING A CURRENT

The discovery (early in the 19th century) that a magnetic field can exert a force on a current-carrying wire led to the speculation, and eventual discovery, that a magnetic field can produce a current. The figure below shows a wire placed between two poles of a (strong) magnet, the ends of the wire being connected to a meter that can detect small currents. If you try this you will find that no current flows while the wire is stationary with respect to the magnet. If the wire is moved up or down through the magnetic field lines, however, the meter will indicate that a current is flowing.

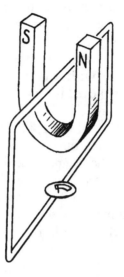

To explain this we will return to the phenomenon of a force on a current-carrying wire. The second figure of this frame represents an enlargement of a section of a wire in a magnetic field, and we imagine that we are able to see a single electron in the wire.

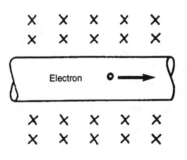

(a) Which way is the electron moving? _____

(b) What is the direction of flow of conventional current? _____

(c) What is the direction of the force on the electron due to the magnetic field?

_____

Answers: (a) to the right;   (b) to the left (Chapter 14, frame 2);   (c) downward (Don't forget that the rule refers to current flow.)

4

The electron in our wire is unable to leave the wire. This creates a force on the wire itself. This is the reason that we observe the force on a current-carrying wire in a magnetic field.

Now consider a different wire. This wire is not connected to a battery or other power source, but it is being moved across a magnetic field. Each electron in the wire is being moved through the field. (See the figure in this frame.) Just as in the previous figure, the electron feels a downward force, and since electrons can move through metal wires, the electron moves downward through the wire. But an electron moving through a wire is an electric current. Therefore, moving the wire relative to the magnetic field has caused a current to be "induced" in the wire. This shows that although the "induction" of a current in a wire appears to be a very different phenomenon than the force that is observed on a current-carrying wire in a magnetic field, they are really caused by the same effect—that a moving electric charge feels a force in a magnetic field.

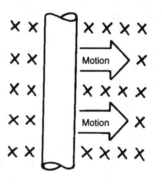

(a) Which way does conventional current flow through the wire in the figure?

_____

(b) What is the effect of moving a wire across magnetic field lines? _____

_____

Answers: (a) up;   (b) to create, or induce, a current

5   A COIL AND A MAGNET

The actual discovery of electromagnetic induction was not made with a horseshoe magnet and a single wire, but with a bar magnet and a coil of wire, as shown in the figure here. When the magnet is held stationary, no current flows through the

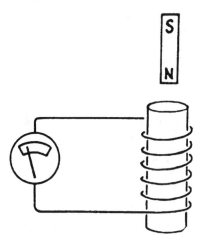

wire of the coil and through the meter, but when the magnet is pushed toward the coil, a current flows.

(a) When the magnet is pulled away from the coil, would you expect current to flow in the same or opposite direction? _____

(b) What would you expect to happen if you added more turns of wire to the coil and moved the magnet in the vicinity of the coil? _____

_____

Answers: (a) the opposite direction (because the relative motion of the wires and magnetic field has been reversed); (b) More current will flow (because more parts of the wire would be cutting across magnetic field lines).

6   LENZ'S LAW

Lenz's law tells us in what direction current will flow when a magnet is moved near a coil.

> The induced current flows in such a direction so that
> its magnetic field opposes the change that produced the current.

When the magnet is at a position shown in the figure of the previous frame, there are relatively few lines of magnetic field through the coil, but when the magnet is pushed closer, the number of lines through the coil increases. The direction of the field represented by these lines is downward (away from the magnet's north pole). A current is induced so as to oppose this increase in the number of lines.

(a) So in what direction will the field caused by the current be? _____

(b) Use the "curled fingers" right-hand rule (from Chapter 15, frames 4 and 6). Which way will current flow across the side of the coil nearer to you in order to produce this field? _____

Answers: (a) upward;   (b) left to right

7

Let's work through another example. Suppose a south pole is pulled away from the top of the coil shown in the figure of frame 5.

(a) Does the magnetic field in the coil increase or decrease as the magnet is pulled away? _____

(b) What is the direction of this field? _____

Now remember that it is this decrease, rather than the existence of the field, that causes the induced current.

(c) In what direction will the induced current seek to establish a field?

_____

(d) How must the induced current flow to get this field? _____

_____

Answers: (a) decrease;   (b) upward;   (c) upward (in order to oppose the decrease); (d) left to right across the front of the coil

---

8   TWO COILS: THE TRANSFORMER

If two coils are wound side-by-side and current is suddenly caused to flow in one coil, a short-lived current will be induced in the other one. This results from the fact that when current starts to flow, a magnetic field is being established where none existed before. Likewise, when an existing current in one of the coils is turned off, a short-lived current is induced in the other.

In the figure of this frame, suppose that someone closes the switch, thereby allowing direct current from the battery to flow in the right coil.

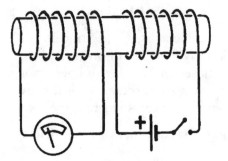

(a) Which direction will current flow in the wire (connected to the battery) that is on your side of the coil? _____

(b) What will be the direction of the magnetic field in the center of the coil?

_____

(c) What will be the direction of the field that the current induced in the left coil seeks to produce? _____

(d) What direction does current flow in the left coil? _____

Answers: (a) downward; (b) left to right; (c) right to left (Lenz's law tells us that the magnetic field of this induced current must oppose the buildup of the left-to-right field.); (d) upward (in the wires on your side)

9

A steady direct current in one coil does not induce a current in the other because then the magnetic field is not changing.

(a) After the switch in the figure is closed for a while, in what direction will current flow in the left coil? _____

(b) When the switch is then opened, stopping the current in the right coil, what is the direction of the current induced in the left one? _____

_____

Answers: (a) It won't flow at all.   (b) down through the wires on your side of the coil (It tends to keep the field from decreasing.)

10

Recall that the magnetic field of an electromagnet can be increased by placing a piece of iron down the center of the coil. The effect in frames 8 and 9 can also be increased by placing a piece of iron through the coils. This figure shows the iron in the form of a ring, which causes the magnetic field to be almost entirely restricted to the centers of the coils, where it will have the most effect. Such a device is called a **transformer**.

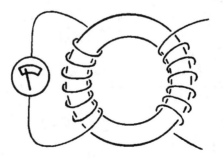

Suppose that the coil on the right is connected to a source of alternating current, so that AC flows in it.

(a) What is the effect of the increasing and decreasing current on the magnetic field produced by the right coil? _____

(b) What is the resultant effect in the left coil? _____

(c) Suppose that instead of AC, DC is caused to flow in the right coil. What will be the resulting effect in the left coil? _____

_____

Answers: (a) It continually varies.   (b) It induces a continually varying current;   (c) No current flows, except when the DC is turned on and off, since the magnetic field doesn't change when DC is flowing.

11 VOLTAGE AND POWER THROUGH A TRANSFORMER

Suppose more turns of wire are placed on the secondary coil (the one at left in the figure of the previous frame) than on the primary (through which current first flows). Each wire of the coils contributes to the induction effect, so the voltage across the ends of the secondary coil will be more than across the primary. In fact, the ratio of voltages is equal to the ratio of turns in the coil:

$$\frac{V_s}{V_p} = \frac{N_s}{N_p}$$

where V represents voltages in the secondary and primary, and N represents the number of turns in each coil. The main function of transformers is to increase or decrease the voltage of alternating current. The large box high on the pole on your street changes the high voltage (perhaps thousands of volts) of the transmission lines to the 110 and 220 volts used in your home.

(a) Suppose the voltage in the lines is 1100 V. Would the number of turns in the secondary to produce 220 volts there be $\frac{1}{5}$ that of the primary, $\frac{1}{10}$, or $\frac{1}{20}$?

_____

(b) If a transformer has 150 turns in the primary and 600 in the secondary, what voltage must be applied across the primary to produce 200 volts across the secondary? _____

Answers: (a) $\frac{1}{5}$;   (b) 50

12

It may appear that the fact that voltage can be increased means that we can increase power output by the use of transformers. This is not the case. Remember that power is the product of voltage and current. If a transformer has twice as many turns in the secondary as in the primary, the voltage is doubled, but the amount of current available to be drawn from the secondary is decreased to one-half. Thus, the product of voltage and current remains the same. We can write the relationship between the currents I and voltages V in the primary and secondary as:

$$I_p V_p = I_s V_s$$

Since the product of current and voltage is equal to power (Chapter 14, frame 20), this equation tells us that the conservation of energy is not violated; there is no more energy flowing from the secondary than flows into the primary.

(a) Two amps at 120 volts flow through the primary of a transformer. If the secondary voltage is 12,000 volts, what current can be drawn from it? _____

_____

(b) One amp at 240 volts flows through the primary of another transformer. If the primary has 500 turns and the secondary 50, what current can be drawn from the secondary? _____

Answers: (a) 0.02 amp; (b) 10 amps ($V_s = 24$; $I_p V_p = I_s V_s$)

13 SELF-INDUCTION

We know that the buildup of magnetic field in a coil causes a current to be induced in the coil. But what about the primary coil in a transformer, the one that is causing the magnetic field in the first place? The field is building up in that coil also. Theoretically, the magnetic field buildup should cause an induced current in that coil, just as it did in the secondary coil. This tendency, called self-induction, does exist.

Refer back to the figure of frame 8. Consider what happens when the switch is closed. Current starts to flow. As soon as it does, the growing magnetic field tries to start a current in the opposite direction. There is, in effect, a "back voltage" placed on the coil that opposes the voltage producing the current. Which of the following would seem to result?

_____ (x) No current flows.

_____ (y) Current increases rapidly to compensate.

_____ (z) Current does not increase as rapidly as expected.

Answer: (z) is what actually happens.

14

Recall that it is the changing of the current, rather than the existence of the current, that causes induction. Because alternating current is by its nature continually changing, a coil of wire impedes its flow. Alternating current that is changing direction with greater frequency causes more induction in a coil, and thus a coil impedes the flow of high-frequency AC more than it does low-frequency AC. Devices called inductors, consisting of wire wrapped around metal cores, are used to allow low frequencies to pass but impede high frequencies.

(a) Will an inductor offer any impedance to the flow of DC? _____

(b) Which of the following frequencies will flow best through an inductor: 60 cycles/second, 300 cycles/second, or 1000 cycles/second? _____

Answers: (a) no; (b) 60 cycles/second

SELF-TEST

1. What will happen when the switch is closed in the circuit of the figure?

   _____

   _____

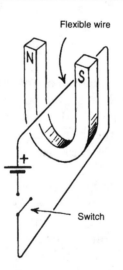

Flexible wire

Switch

2. Suppose a current runs down a wire from the top to the bottom of this page, and that a magnetic field runs through the page from the other side toward you. What will be the direction of the force on the wire?_____

   _____

3. Suppose a wire carries a current from east to west along the equator, where the magnetic field is from south to north (geographic directions—recall that the magnetic poles are near the "wrong" geographic poles). What is the direction of the force on the wire? _____

4. Refer to the first figure in frame 3. What will happen if the magnet is quickly moved upward (while the wire remains stationary)? _____

5. Suppose the north pole of a magnet is pulled away from the *lower* end of a coil that is wound like that of the figure in frame 5. In which direction will the induced current flow through the wires that can be seen on the side of the coil facing you? _____

6. Why is an iron core included in transformers? _____

7. Why can a transformer not be used to transmit direct current? _____

8. Can a transformer be used to increase the power output of an electric circuit? Why or why not? _____

9. A transformer has 500 turns in the primary and 3000 in the secondary. If 120 volts AC is put across the primary, what voltage will appear across the secondary? _____

10. If 2.0 amps flow through the primary of the transformer of question 9, what is the maximum current that can be drawn from the secondary? _____

ANSWERS

1. The wire will move (downward) between the poles of the magnet due to the force on the current-carrying wire. (frames 1, 2)

2. toward the left (Use the right-hand rule.) (frame 2)

3. downward, toward the ground (right-hand rule) (frame 2)

4. A current will be induced in the wire. (frame 3)

5. from left to right (As the magnet is pulled away, the field—which is upward—decreases in the coil. The induced current opposes this decrease. Thus, it is such as to cause an upward magnetic field. The curled fingers right-hand rule of the last chapter tells us that it must move across the visible wires from left to right.) (frame 6)

6. The iron core increases the magnetic effect by adding its own magnetic field. (frame 10)

7. A change in magnetic field through the secondary is necessary to induce a current in it. A direct current will produce a constant field rather than a changing one. (frame 9)

8. No. Although voltage can be made to increase, current will be decreased in the same proportion. The law of conservation of energy is not violated. (frame 12)

9. 720 volts (frame 11)

10. 0.33 amp (frame 12)

17 Electromagnetic Waves

Prerequisites: Chapters 13 (frames 6–8), 14–16

The link between electricity and magnetism on the one hand and light on the other is not an obvious one. Sure, we all flick an electric switch to turn on a light, but it is electromagnetic fields that provide the real link between the two branches of physics. We will see that visible light can be considered as waves that originate in electric and magnetic fields. The details of this connection will be saved until the next chapter; right now we will spend most of our time on another type of electromagnetic wave—the radio wave.

OBJECTIVES

After completing this chapter, you will be able to

- describe how a changing electric field is generated by an antenna;
- specify the condition of an antenna when its electric field is at a maximum and when its magnetic field is at a maximum;
- identify a factor important in the construction of a transmitting radio antenna that is related to the frequency of the wave;
- differentiate between the processes of AM and FM radio transmission;
- identify, in order of frequency, a number of portions of the electromagnetic spectrum;
- describe two methods of detecting an electromagnetic wave;
- relate the number on a radio dial to the radio wave that is received;
- relate light to electromagnetic waves.

1 PRODUCTION OF RADIO WAVES

Consider two wires connected to a source of alternating voltage as in figure A below. The source causes electrons to move up into the upper wire, then down into the lower, then back up, and so on. Figure A shows the device at the instant when the top of it is positive and the bottom is negative. At this instant, there is an electric field (represented by the arrows) in the region immediately surrounding the device. The field points downward.

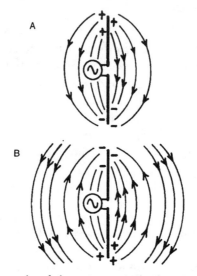

A moment later, the ends of the wire switch electric polarity, and the situation of figure B occurs.

(a) In what direction does the electric field near the wire point now? _____

(b) Has the previous downwardly directed field disappeared? (Refer to the figure.) _____

(c) Where has it gone? _____

Answers: (a) upward; (b) no; (c) It has moved farther from the device.

2

The field continues to spread out from the source so that an electron located some distance from it would feel a force first upward, then downward, then upward, and so on. (Recall that the electron, having a negative charge, feels a force in a direction opposite to the direction of the field.) An electric wave has been sent through the space between the device and the electron that feels the force. In fact, if this electron is one of the many located in a receiving antenna (which is very similar to the device illustrated in frame 1 except that a current detector replaces the AC source), the electric wave can be detected at great distances. The antenna on your car works in this way, detecting the electric field wave transmitted from the radio station.

But the story does not stop here. The figure below shows the same device—a transmitting antenna—during the time when the electrons are in motion from the bottom to the top of the wires. Since a current is flowing in the wires, a magnetic field surrounds them. The field is shown into the paper on the left side and out of the paper on the right because the conventional current is downward and the right-hand rule gives this as the correct direction of the field.

Current flows down the wire until the bottom is positive and top negative (at which time the electric field is at a maximum), but then it starts flowing upward, producing a magnetic field in a direction opposite to that of the figure in this frame. Just as the electric field spreads out from its source after being produced, a magnetic field that is detected at some distance from the transmitting antenna will point first in one direction and then in the other. It will alternate direction with the same frequency as the electric field and the same frequency as the alternating current in the antenna.

We saw in the previous chapter that a changing magnetic field causes a current to flow in a coil of wire. Indeed, a coil that has its ends connected to a device that responds to the small currents induced in the coil is effective in detecting the magnetic wave described here. Your home radio very likely has a coil of wire inside (or an external coil) that serves as its antenna to detect magnetic waves transmitted by radio stations.

Neither an electric wave nor a magnetic wave can exist without the other; their combination is called an **electromagnetic wave**.

(a) What two components make up an electromagnetic wave? _____

_____

(b) Describe what is happening in a transmitting antenna when it produces a maximum electric field. _____

_____

(c) When does it produce the maximum magnetic field? _____

_____

(d) What is necessary to induce a current in a coil of wire? _____

_____

Answers: (a) electric wave and magnetic wave; (b) One end is positive and the other is negative. (c) When the current in it is at a maximum. (d) a changing magnetic field

3

One might think that the frequency with which electromagnetic waves are sent out depends only upon the frequency of the AC source. In practice, however, the length of the antenna is important in getting maximum wave amplitude. Just as a strip of metal (like a saw blade) that is clamped at one end has a natural frequency at which it will vibrate back and forth, there is a natural frequency with which electrons alternate back and forth in an antenna. The value of this frequency is determined by the length of the antenna. To make an antenna that will transmit a wave of maximum amplitude, it must be constructed such that its natural electron-vibration frequency corresponds to the AC frequency.* Antennas such as the one shown in frame 2 send out radio waves that may have frequencies as low as a few cycles per second up to about a billion cycles per second. This is a far greater range than the radio waves we normally think of. AM radio uses a range from about 530 to about 1600 kilocycles per second, while FM radio frequencies range from about 88 to 108 million cycles per second. We use the remainder of the radio band for such purposes as citizen's band radio, television, short wave radio, and microwave ovens.

(a) What affects the amplitude of the transmitted wave besides the voltage of the AC source? _____

(b) Name four uses of various parts of the radio section of the electromagnetic spectrum. _____

_____

Answers: (a) antenna length; (b) AM radio, FM radio, television, citizen's band radio, shortwave radio, microwave

4

If the electromagnetic wave described in frame 1 were represented graphically, it would look like wave A in the figure of this frame. This graph shows us that the wave is a smooth, regular wave. In AM radio, a signal is put on the wave by amplitude modulation. For example, suppose wave B represents a sound wave

*If you studied Chapter 11, you will recognize such frequency matching as the condition necessary for "resonance."

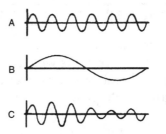

that the local disc jockey wishes to transmit. This wave is used to "modulate" the basic radio wave (called the carrier wave) so that the transmitted wave looks like wave C. The amplitude of the sound wave has been used to control—to "modulate"—the amplitude of the carrier wave. Look carefully to see the similarity of pattern between graphs B and C of the figure.

In tuning your radio receiver, you adjust the frequency of your radio to the frequency of the carrier wave (so that only the desired station is received), and your radio set separates the sound wave from the carrier wave and reproduces the former as sound.

What does "amplitude modulation" do to the radio wave? _____

_____

Answer: It changes the amplitude of the carrier wave according to the sound frequency.

5

FM stands for frequency modulation. In FM radio, it is not the amplitude of the wave that is modulated, but the frequency. Wave A in the figure of this frame is a carrier wave (unmodulated). Wave B of the figure is a sound wave that is used to modulate the carrier wave. The modulated wave, C, has been changed in frequency so that it carries the sound signal. Although the electronic details of the transmitter and receiver differ, the principle is the same as for AM radio.

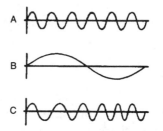

What is the basic difference between AM and FM radio signals? _____

_____

Answer: In AM the amplitude of the carrier wave is modified; in FM the frequency is modified.

6 THE ELECTROMAGNETIC SPECTRUM

As great a range as they cover in frequency, radio waves are only part of the electromagnetic spectrum. As we advance to frequencies higher than about a billion cycles per second, we come to the infrared part of the spectrum. Infrared waves are basically the same type as radio waves, but they have many properties that are different from those of radio waves. Infrared waves are perceived by the body as heat radiation. The figure here shows that the region of next highest frequency beyond infrared waves is that of visible light. It was mentioned at the beginning of this chapter that light and electromagnetism have a link. You can see from this figure that visible light is part of the spectrum of electromagnetic waves.

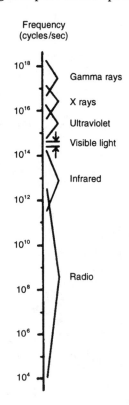

I said earlier that the size of the transmitting antenna is related to the frequency of the wave produced. Infrared waves cannot be produced by wire antennas because these antennas cannot be made short enough for the high frequency of infrared. An extremely short antenna would be needed—in fact, about the size of an atom. Vibration of atoms is what produces infrared radiation. And, to produce light, the vibration must occur within the atom itself. The production of light and its properties are the subjects that I will take up in the next few chapters.

(a) Refer to the figure. What types of waves have greater frequencies than light?

_____

(b) What is the highest frequency for which antennas can be made (approximately)? _____

Answers: (a) ultraviolet, X rays, gamma rays;   (b) 10^{12} to 10^{13} cycles/second

SELF-TEST

1. On the illustration of a transmitting radio antenna here, show the arrangement of electric charges when a maximum downward electric field is being transmitted.

2. What is the condition of the electrons in the transmitting antenna when maximum magnetic field is being transmitted? _____

   _____

   _____

3. How can the electric field portion of an electromagnetic wave be detected?

   _____

   _____

4. How can the magnetic field portion of an electromagnetic wave be detected?

   _____

   _____

5. Besides the frequency of the alternating current source, what else is important in determining the frequency of a transmitted electromagnetic wave?

   _____

6. Choose some number on an AM radio dial and explain what is meant by that number. _____

_____

_____

7. What is meant by amplitude modulation? _____

_____

_____

8. Explain what frequency modulation means. _____

_____

_____

9. List, in order of frequency, from low to high, at least four portions of the electromagnetic spectrum. _____

_____

_____

10. What basic connection is there between light and electromagnetism? _____

_____

_____

ANSWERS

1. Positive charges are at the top of the antenna and negative charges at the bottom. (frame 1)

2. A maximum current is flowing, so the electrons have maximum speed up or down the antenna. The electric current produces the magnetic field. (frame 2)

3. It can be detected by an antenna that consists of a long wire attached to a device that is sensitive to changes in the flow of electric current. (frame 2)

4. The magnetic wave is best detected by placing a loop of wire (with its ends hooked to a current detector) in the path of the wave. The changing magnetic field causes a current in the loop. (frame 2)

5. the length of the transmitting antenna (frame 3)

6. For example, "900" on the dial means that when you are tuned to that number, your radio is detecting a carrier frequency of 900,000 cycles/second. (Some dials may show this as "90" or "9.") (frame 3)

7. In amplitude modulation, the amplitude, or intensity, of the carrier wave is changed according to the frequency of the sound being transmitted. (frame 4)

8. In frequency modulation, it is the frequency of the carrier wave that is changed, or modulated, according to the frequency of the sound being transmitted. (frame 5)

9. The ones listed in this chapter, from low to high frequency, are as follows: radio, infrared, visible, ultraviolet, X rays, and gamma rays. (frame 6 and the figure there)

10. Light acts as an electromagnetic wave and therefore is part of the electromagnetic spectrum. (frame 6)

<u>18</u> Light: Wave or Particle?

No prerequisites, but Chapters 10–12 recommended
The nature of light has been a puzzle down through the ages. Early theories included the idea that the eye sends out particles that go to the thing seen. In newspaper cartoons even today you sometimes see little arrows pointing from a character's eye to an object to illustrate that the character is looking at the object. Eventually people asked the question, if this is the nature of light, why can't we see in the dark? As we will see, however, the real conflict historically has been between theories that demand much more serious treatment than this primitive one.

OBJECTIVES

After completing this chapter, you will be able to

- specify why Galileo failed to measure the speed of light with his experiment;
- describe how Roemer used the period of one of Jupiter's moons to measure the speed of light;
- explain how Michelson measured the speed of light on earth;
- state the speed of light in metric and British units;
- specify why Newton clung to a particle theory of light;
- identify the scientist who discovered that light definitely acts as an electromagnetic wave and state his method of discovery;
- interpret the chart of the electromagnetic spectrum;
- interpret the observations of the photoelectric effect based on wave theory;
- explain why Michelson and Morley expected to be able to detect the ether;
- state evidence for and against both the wave and the particle nature of light.

1 GALILEO AND THE SPEED OF LIGHT

One of the problems in analyzing light is simply to measure its speed. In about the year 1600, Galileo Galilei tried to do this. He stationed himself on a hill and used a lantern to send a flash of light to a friend on another hill a few miles away. When the friend saw the flash from Galileo's lantern, he would answer with a flash from his. All Galileo would have to do was to measure the time from when he exposed his lantern to when he saw the return flash and then subtract his friend's reaction time. All Galileo could conclude was that light moves faster than he could measure. He could not even rule out the possibility that it moves instantly from place to place—with infinite velocity.

Which of the following could explain Galileo's results?

_____ (x) Measurement of time had not progressed enough to measure the speed of light.

_____ (y) His lanterns were too close together.

_____ (z) His lanterns were too far apart.

Answer: (x) (In fact, it was Galileo who invented the pendulum clock. Answer (y) may be considered a possible answer, but there could be no hills far enough apart to allow the measurement without the ability to measure extremely short times.)

2 ROEMER'S MEASUREMENT

The first successful measurement of the speed of light was reported by Olaus Roemer in 1675. His method was based on the astronomical observation that Io, the innermost moon of the planet Jupiter, has a period of revolution around the planet of 42.5 hours.* Once Roemer observed it crossing in front of Jupiter, he could predict that it would again eclipse Jupiter 42.5 hours later. And every 42.5 hours after that. In fact, he could predict accurately at what time of night Io would eclipse Jupiter 3 months or 6 months later.

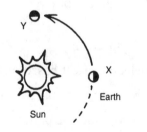

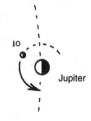

---

*You may be interested to know that it was Galileo who discovered that Jupiter has moons—the first, other than our own, ever seen.

Consider the figure. When the earth is at point X, Roemer observes Io eclipsing Jupiter. He does some calculations and predicts that 3 months later this should happen again at a certain time of night. However, when he makes his observations 3 months later, he finds the measurement is wrong. What is the most likely explanation of this?

_____ (x) The period of Io changed.

_____ (y) The earth is farther from Jupiter.

_____ (z) Jupiter is in a different position.

Answer: (y) (The period of a moon doesn't change measurably in 3 months, and Jupiter does not move far enough in that time to make a difference.)

3 When the earth got to position Y (in the figure of frame 2) Roemer found that the eclipse happened 500 seconds later than predicted. This unexpected result is not due to an irregularity in the orbit of Io, but is due to the fact that light from Jupiter requires 500 seconds longer to reach earth at position Y than it did when the earth was at X.

To calculate the speed of light, all Roemer had to do was divide the earth-to-sun distance by the 500 seconds. However, he did not know the earth-to-sun distance as accurately as we do now, so the value he obtained was not as accurate as we can calculate. His observation and calculation, however, formed the first successful measurement of the speed of light. Let's use our known distance from the sun ($1.5 \cdot 10^{11}$ meters)* and divide by the time (500 seconds) to find the speed of light. _____

Answer: $3 \cdot 10^8$ m/s (or 186,000 miles/s)

4 MICHELSON'S METHOD

Late in the nineteenth century, Albert Michelson devised an earthbound method to measure the speed of light. The figure on the next page shows light from a source (upper left) hitting an eight-sided mirror, bouncing over to a regular mirror (M), bouncing back to hit another side of the eight-sided mirror, and finally bouncing into Michelson's eye at the bottom. Now suppose that Michelson rotates the octagonal mirror clockwise. Flashes of light will now be sent down to mirror M. When the rotating mirror is at the position shown in the figure, light will leave it in the correct direction to hit M and be returned again. But by that time, side Z (where it should hit) will have moved around a little, so that the light flash will not be bounced into Michelson's eye. So Michelson speeds up the

*This equivalent to 93,000,000 miles, the average radius of the earth's orbit.

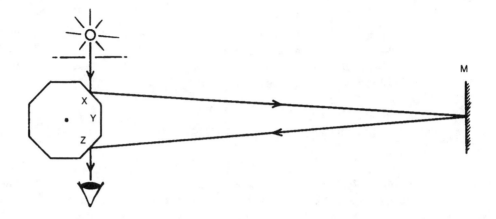

rotation. He speeds it up until, during the time that the light flash is going from X to M and back, side Y has moved over to replace Z in its position. The light then reflects from Y to the waiting eye. All that Michelson needed to know was the rotational speed of the mirror (from which he could calculate time) and the distance to mirror M.* Distance divided by time yielded for Michelson the speed of light.

The advantage of Michelson's calculation over that of Roemer was that Michelson could measure his variables more accurately, and he wasn't dependent on astronomical measurements.

If Michelson had used a 12-sided mirror rather than an eight-sided one, would he have to rotate it faster or could he have rotated it more slowly?

Answer: more slowly (Only $\frac{1}{12}$ of a rotation would have been required while the light was traveling.)

5   HUYGENS' PRINCIPLE

Early experimenters showed that light does not travel at an infinite speed, and that its speed can be measured. But what is light? The ancient view was that it is a stream of particles, or corpuscles, that leave the object being viewed and travel to the eye. In the mid-1600s, Robert Hooke, an English physicist, proposed that light may be a wave. This theory was then improved by Christian Huygens, who explained that light waves leaving a source could be considered to be the result of a number of tiny wavelets, each emitted by each point on the wave. Thus, light leaving point A in the figure on the next page consists of little wavelets that at

---

*The distance to mirror M was more than 20 miles in the original experiment.

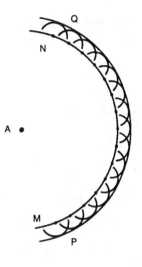

some time later appear as the wave from M to N. To see how the wave spreads from there, suppose that each point on the wave MN emits a little wavelet. These wavelets spread out so that a little later they will appear as the many wavelets that combine to form a single wave stretching from P to Q. (In Chapter 12 we saw that little waves *interfere* with one another, and superimpose to form the resulting wave.) This method of considering the progression of waves is named after the man who suggested it and is called Huygens' principle.

(a) The figure in this frame illustrates waves in two dimensions. How do light waves differ from this? _____

(b) In two dimensions, each wavelet forms part of a circle. What is formed in three dimensions? _____

Answers: (a) They spread in all directions and in three dimensions.   (b) part of a sphere

6 │ NEWTON'S PARTICLES

As you probably know, Isaac Newton is one of the giants—perhaps *the* giant—of classical physics. Newton was a contemporary of Robert Hooke, but he could not accept Hooke's wave theory of light. Newton realized that if light is a wave, it should bend around corners. Newton hadn't observed this, so he concluded that light could not be a wave.

Newton had done experiments with light and had observed that it breaks into colors when it passes through a prism, but that the separated colors themselves can no longer be broken further into more colors. His theory was that each color corresponds to a certain type of particle, or corpuscle. White light, according to this theory, is composed of particles of all types. Each type passes through the glass of a prism at a different speed, and this causes each to bend a different amount. Thus, the light separates into different colors. He also observed that when all of the colors are bent back together again, white light results. This fit the theory well.

Since Newton had such a great reputation, the fact that he favored a particle theory of light may have resulted in slowing the progress of work toward finding the true nature of light. After all, if the master says that light is a particle, why experiment and theorize in any other direction? Which of the following did Newton use as evidence that light acts as a particle rather than as a wave?

_____   (w) Light doesn't seem to bend around corners.

_____   (x) The speed of light can be explained by considering it a particle.

_____   (y) Light is emitted by hot objects.

_____   (z) The separation of light into colors can be explained with a particle model.

Answer: (w) and (z)

7   ELECTROMAGNETIC WAVES

In the early 1800s Thomas Young performed an experiment that showed conclusively that light has wave characteristics. (If you have studied Chapter 12, you may want to jump ahead to frames 1 through 4 of Chapter 22 to see the nature of this experiment.) But it was not until the middle of the 1800s that the nature of the wave was discovered. James Clerk Maxwell was doing theoretical work on electricity and magnetism at the time. This work led him to predict the possibility of the existence of electromagnetic waves—waves given off when electrical charges vibrate in a wire. Maxwell formulated a set of equations that, when the waves were observed 30 years later, were found to fit their behavior quite well. In the course of his work, he did a calculation of the predicted speed of the electromagnetic waves. The speed his calculation showed was $3.0 \cdot 10^8$ meters per second. He realized, of course, that this was the same as the speed of light, and since this was not likely to be just a coincidence, he concluded that light is a form of electromagnetic wave. We know now that all electromagnetic waves move at this speed.

(a) Who first showed experimentally that light acts as a wave? _____

_____

(b) Who first concluded that light was an electromagnetic wave? _____

_____

Answers: (a) Thomas Young;   (b) James Clerk Maxwell

8

You may know that the frequency of sound waves determines the pitch of the sound we hear. In the case of light, the frequency of the light wave determines the color of the light we see. The waves in the visible spectrum have frequencies between about $4 \cdot 10^{14}$ cycles/second and $7.5 \cdot 10^{14}$ cycles/second. (This is from 400,000,000,000,000 to 750,000,000,000,000.)* Red light has the lowest frequency. Then comes orange, yellow, green, blue, indigo, and finally violet at the highest frequency.

The electromagnetic spectrum stretches far beyond the limits of visible light, with frequencies much higher and much lower. The figure here illustrates this range. Just beyond the violet, we find ultraviolet, and the highest frequency waves are gamma rays. At the low end, the spectrum ranges from infrared to radio waves.

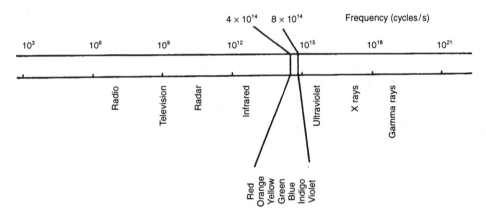

(a) Refer to the figure to determine approximately what frequency electromagnetic wave is used for television. _____

(b) What is the speed with which X rays travel? _____

Answers: (a) about 10^9 cycles/s;   (b) $3 \cdot 10^8$ m/s (Or 186,000 miles/s. All electromagnetic waves travel at the same speed; see frame 7.)

9 # THE PHOTOELECTRIC EFFECT

Maxwell's work with the theory of electromagnetic waves may seem to have solved the problem of the nature of light, but at least one major problem remained. There was one experiment—the photoelectric effect—that could not be explained by considering light as a wave. Suppose we set up an electric circuit as shown in the next figure. In this circuit the negative terminal of a battery has been connected to a piece of sodium metal. The positive terminal of the battery is

---

*See Appendix I for a discussion of powers-of-ten notation.

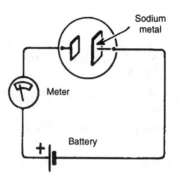

connected through a meter that measures electric current to another piece of metal a centimeter or so from the sodium. Both of these metal plates, as they are called, are enclosed in a sealed glass tube from which the air has been removed. When this experiment is performed in the dark, no current flows from the battery, and the electric meter shows a reading of zero. If light is shined onto the sodium plate, electric current will flow and register on the meter. If you block the light that hits the sodium plate, the electric current stops. And, as one might suspect, when the amount of light that hits this negative plate is increased, the amount of current is increased.

But this is still not the whole story. It turns out that if we try various colors of light, we find that violet and blue light are able to cause current to flow, but colors toward the red end of the spectrum do not result in electric current when they are shined on the sodium. Somehow, the frequency of the light is important, with higher frequency light causing current and lower frequency light unable to do so.

In Chapter 14 we saw that electric current in a wire consists of electrons flowing from the negative end of the wire to the positive end. In the circuit of interest here, no electrons were able to flow when no light hit the sodium because the electrons were not released from the negative plate so that they would move through the vacuum to the positive plate. Somehow the light is able to bump electrons out of the negative plate. Then once they are free from that plate, they are attracted to the positive plate and they continue on through the meter to the positive terminal of the battery.

(a) What is needed to cause current to flow in the circuit shown in this frame?

_____

(b) What component of the circuit emits electrons when light hits it?

_____

Answers: (a) high-frequency light;   (b) sodium metal

10

A summary of the observations of the photoelectric effect follows:

1. Current flows as soon as light hits the negative plate. No "buildup time" is needed, no matter how dim the light.

2. High-frequency light causes electrons to be emitted from the sodium, but low-frequency light does not. (The critical frequency below which electrons will not be emitted depends upon the metal used as the negative plate. The vast majority of metals require light of frequencies well above the visible region of the spectrum.)

3. Once the frequency of light is great enough to cause electron emission, the energy of the emitted electrons does not depend upon the brightness of the light. It depends upon the light's frequency, higher frequency light causing higher energy electrons to be emitted.

4. The amount of current that flows depends upon the brightness of the light.

(a) Which will cause more current to flow, a brighter light or a higher frequency light? _____

(b) Which would produce emission of higher energy electrons, blue light or violet light? _____

Answers: (a) brighter; (b) violet

11 ELECTROMAGNETIC THEORY AND THE PHOTOELECTRIC EFFECT

Because of the success of Maxwell's electromagnetic theory, at the end of the 1800s light was considered to be wavelike in nature. Let us see how well this theory explains the photoelectric effect. Read the statements below—all of which are true—and relate each of them to observations in the last frame. Indicate which facts support the wave theory as an explanation for the photoelectric effect.

_____ (w) The energy contained in electromagnetic waves and the amount of that energy that would strike an individual electron can be calculated. Such a calculation shows that an electron could indeed gain enough energy to break free from sodium, but only after light was shined on it for some hours (the amount of time depending upon the brightness of the light).

_____ (x) Since light waves carry energy whether they are of high or low frequency, the frequency of light should have no effect on whether or not electrons are emitted from the metal.

_____ (y) The frequency of the light wave should not affect the energy of emitted electrons. If anything, one might expect the amplitude of the waves (rather than the frequency) to affect the energy of the emitted electrons.

_____ (z) Brighter light carries more energy and might be expected to eject more electrons.

Answer: (z)

Since the ability to explain only one in four of the observations is not nearly enough, the photoelectric effect was a major roadblock in the way of total acceptance of the wave theory of light.

12 THE ETHER

Another problem with the electromagnetic theory arose over the question of the ether. This ether (sometimes spelled "aether") has nothing to do with the ether gas we use today in medicine. It was the medium through which electromagnetic waves were expected to pass. Every other wave has some medium through which it passes. Water waves require a water surface; waves going down a stretched spring obviously require a spring; and sound waves likewise require a material (usually air) for their propagation. The question arose as to what medium propagates light waves. This medium must possess some unusual characteristics:

> It must exist throughout space, all the way to the stars. The earth, in its journey around the sun, would be continually moving through the ether, but the earth apparently passes through it without friction. No material known would allow such motion through it, but the ether must somehow be able to.
> Since light passes through some solids, such as glass, the ether must actually exist inside these transparent substances.

The ether, then, must not be a *material* substance at all. It must not be made up of atoms and molecules.

(a) Why must the ether extend to the stars? _____

(b) Why must the ether not be a material substance? _____

_____

Answers: (a) Light reaches us from the stars. (b) It must permeate material such as glass and water. In addition, the earth must pass through it without friction.

13 THE MICHELSON–MORLEY EXPERIMENT

The speed of sound *relative to the air* is 1100 feet/second. Consider a man at the center of a railway flatcar, as shown in the figure. If the car is standing still and

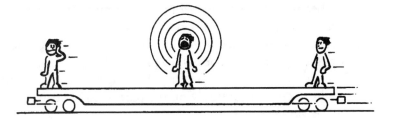

there is no wind, a shout from the man will be heard at the same time by a person at each end of the car.

But what happens when the car is in motion at, say, 60 ft/s (about 40 mi/hr)? In this case air is swishing by the shouting man at 60 ft/s. When a person shouts, she or he does not *throw* the sound as one throws a baseball; the person simply disturbs the air and the disturbance moves away through the air at 1100 ft/s. But if the air is moving toward the shouter at 60 ft/s, then the sound actually moves away from the man toward the front of the railway car at 1040 ft/s. And relative to the car, it moves toward the person at the rear at the rate of 1160 ft/s. If the person at the rear did not take into account that the car was moving, what would she measure the speed of sound to be? _____

Answer: 1160 ft/s

14

Now let's consider light and the ether. The velocity of light is $3 \cdot 10^8$ m/s (or 186,000 mi/s). The earth, however, must be moving through the ether in much the same way that the railway car is moving through the air. If this is the case, the speed of light should be slightly more or less, depending upon which direction the light is traveling through the ether. The earth's speed around the sun is about $3 \cdot 10^4$ m/s (20 mi/s), very slow compared to the speed of light. Thus, the difference in the speed of light due to the motion of the earth should be very difficult to detect.

In 1881, Albert Michelson and Edward Morley decided to try to detect the ether. To perform this very difficult measurement, Michelson and Morley used the experimental setup illustrated in the figure here.

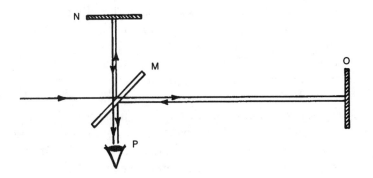

A beam of light comes from the left and hits mirror M, which is only half-silvered so that half of the light passes through while the other half is reflected. The reflected light goes up to mirror N and bounces back toward M. The light that passes through M goes on to mirror O and likewise bounces back toward M. When each of these light beams hits M the second time, it is again divided, but we want to concentrate only on the part of each beam that ends up going toward the observing eye at P.

(a) What happens when light hits a half-silvered mirror? _____

_____

(b) What path did the light follow in getting from the source to the eye in the Michelson–Morley experiment? _____

_____

Answers: (a) Half goes on through and half is reflected.   (b) to mirror M, then half to N and half to O, then both halves back to M, then to the eye

---

15   Suppose each of the beams of light travels exactly the same distance from the time it leaves M until it returns to M. In that case, a crest of the wave that went one direction will exactly meet the crest of the wave going the other way. Likewise, troughs of the two waves will meet at M. In Chapter 12 we saw that this caused constructive interference. If you did not study that chapter, suffice it to say that when crest meets crest and trough meets trough, the light that went one way will add to and reinforce the light that went the other way.

On the other hand, if the light travels just slightly slower along one path, a crest from one direction will land on a trough from the other, resulting in destructive interference. This means that the two beams of light will cancel one another out.

In actual practice, Michelson and Morley did not have to set the paths to exactly the same length. They set the apparatus up so that they saw a pattern of interference when they made an observation with the apparatus at a fixed position. Then they rotated the entire setup by 90 degrees and they looked for a change in the pattern. Why did they expect to see a change? Suppose that as they originally started the experiment, the ether was flowing past the earth in the direction from mirror N to mirror M. The light going from M to N was thus fighting against the "ether wind." In going back to M, it was moving with the ether wind. Light making the trip from M to O and back, on the other hand, was continually moving in a direction across the wind. Analysis of such motions predicts that the speed across the wind will be greater than the average speed upwind and downwind. Now the experimenters rotate the apparatus. Light that was once going across the wind now travels up and down, and light that traveled up and down is now going across. Thus, no matter how the two distances compare, there should be a change in the interference pattern when the apparatus is rotated. To make a long story less long, no change was seen. The experiment was tried again and again, and calculations (which indicated that such a change should show up) were checked and rechecked. The result was the same—no change in the interference pattern. Two conclusions were possible: (1) maybe the ether travels along with the earth so that it causes no ether wind, or (2) there is no ether. The first of these is ridiculous—we found out long ago that the earth is not at the center of the universe. We can't expect ether to follow the earth around. We have to conclude that there is no ether. And we are back to the problem of what medium light travels in if light is a wave. The only answer that we can give, even today, is that electromagnetic radiation is a vibration of the electromagnetic field, which is not a material substance. Light is the "force-signal" sent by an electric charge through this field, and there is no medium needed for electromagnetic waves except the field that the wave establishes.

(a) Why was it not necessary for each light path to be exactly the same length in the experiment? _____

_____

(b) What was the overall purpose of the Michelson–Morley experiment?

_____

Answers: (a) Michelson and Morley were looking for a change as they rotated the apparatus. (b) to detect the ether (or the earth's motion through the ether)

16 SUMMARY

Is light a wave or a particle? We have examined some of the evidence pertaining to this question and have concluded that it doesn't seem to be either. Maxwell's electromagnetic theory seems to explain most of the observations very well, but it falls short on the experiment concerning the photoelectric effect. In addition, we have to conclude that there is no substance that carries the wave.

But why do we expect light to be *either* a wave or a particle? We expect this because we cannot imagine anything else the light could be. But that does not mean that light cannot be something else. There is no question but that light has wave aspects—it exhibits the phenomenon of interference, and that cannot be explained except with a wave explanation. But it also exhibits particle-type effects in the photoelectric effect—it acts as if particles are knocking electrons out of a metal. Perhaps we cannot say that it is a wave or that it is a particle, but must live with the fact that whatever it is, it sometimes acts like a wave and sometimes acts like a particle. We will see in the next chapter that there is a theory that, at least in part, describes light by uniting these two aspects of nature.

SELF-TEST

The questions below will test your understanding of this chapter. Use a separate sheet of paper for your diagrams or calculations. Compare your answers with the answers provided following the test.

1. What did Galileo conclude concerning the speed of light? _____

   _____

2. What causes the apparent variation in the period of the innermost moon of Jupiter? _____

   _____

   _____

3. In Michelson's method for measuring the speed of light, how did he determine how long the light took to go to the distant mirror and return to his eight-sided mirror? _____

   _____

4. What is the speed of light? _____

5. Did Isaac Newton favor the wave or particle theory of light? _____

   _____

6. What did Newton propose as the cause of light having various colors?

   _____

   _____

7. What was surprising about the speed Maxwell predicted for electromagnetic waves? _____

8. (a) What spectral color has the highest frequency? _____

   (b) What type waves have a frequency just higher than this? _____

9. Will light of high frequency or of low frequency be more likely to cause photoelectric emission of electrons? _____

10. What does wave theory predict concerning the time required for electrons to be emitted once light is shined on a metal? _____

    _____

11. What effect does wave theory predict that the frequency of light will have on the emission of electrons from metals? _____

    _____

12. The Michelson–Morley experiment was an attempt to determine the earth's speed through what? _____

13. What was concluded from the Michelson–Morley experiment? _____

_____

ANSWERS

If your answers do not agree with those given below, review the frames indicated in parentheses before you go on to the next chapter.

1. Light is faster than he could measure, and perhaps infinitely fast. (frame 1)

2. The apparent variation results from the fact that sometimes the earth is farther from Jupiter than at other times, and it takes light longer to get to us when we are farther away. (frames 2, 3)

3. This time was equal to the time for his mirror to rotate through $\frac{1}{8}$ turn. He had to know the frequency of rotation of the mirror. (frame 4)

4. $3 \cdot 10^8$ m/s or 186,000 mi/s (frame 3)

5. the particle theory (frame 6)

6. Newton proposed that each color corresponds to a certain type of particle. (frame 6)

7. It was the same as the speed of light. (frame 7)

8. (a) violet;   (b) ultraviolet (frame 8 and the figure there)

9. high (frame 11)

10. Some time should be required while the electrons build up energy. (frame 11)

11. It should not affect the emission of electrons. (frames 11)

12. ether (frames 14, 15)

13. There is no evidence for the existence of an ether. (frame 15)

$\underline{19}$ The Quantum Nature of Light

Prerequisites: Chapter 5 and frames 8–10 of Chapter 18
In the last chapter we saw that many aspects of the behavior of light can be understood when you consider it to be a wave. Some of its behavior, however, demands a particle explanation. In this chapter we will discuss more of these phenomena and will present a view of light that—at least partly—unites the wave and particle nature of light.

OBJECTIVES

After completing this chapter, you will be able to

- relate blackbody radiation to the temperature of the object;
- identify the problem concerning the relationship between Maxwell's electromagnetic theory and blackbody radiation;
- specify Planck's contribution to the study of the nature of light;
- relate Planck's formula to the wave and quantum theories of light;
- specify Einstein's contribution to the theory of light;
- interpret the photoelectric effect with regard to the quantum theory;
- relate the Bohr model of the atom to the quantum theory of light;
- determine the amount of energy emitted when an atom de-excites, or the energy needed to excite;
- differentiate between excited and stable states in an atom;
- relate the frequencies of emitted photons to the use of spectra in identifying elements;
- name at least two ways to raise atoms to excited states;

- differentiate between emission spectra, absorption spectra, and the continuous spectrum;
- explain the difference in functioning of an incandescent and a fluorescent lamp;
- compare what happens when atoms in a solid are excited with what happens when those in a gas are excited;
- distinguish between fluorescent and phosphorescent materials;
- explain how "black light" posters work;
- state the difference between laser light and regular light;
- specify the derivation of the word "laser."

1 BLACKBODY RADIATION AND ELECTROMAGNETIC THEORY

If you turn on a burner of an electric stove to its lowest setting, you can feel heat radiated from the burner even though it does not glow visibly. As you adjust the burner to a hotter setting, you can feel more heat radiated and you begin to see the burner glowing red. With a yet higher setting, the burner goes from red to bright yellowish-red. The figure is a graph of the intensity of radiation emitted by an object at various temperatures. The line that corresponds to 6000°C shows that most of the radiation from an object at this temperature is emitted at a higher frequency than radiation from objects at 4000°C and 3000°C.

The heat you feel radiated near the stove burner is infrared radiation, which is of lower frequency than visible light. When the burner is set at low, it is not hot enough to emit visible light. As the temperature is increased, however, it starts emitting light of higher and higher frequency until red light is emitted. As it continues getting hotter, it emits more light toward the higher frequency end of the

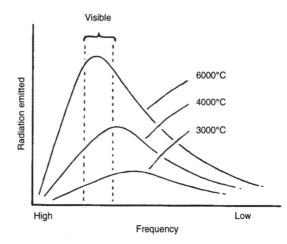

spectrum, causing the appearance of an orangeish- or yellowish-red. If you were able to continue increasing its temperature, the burner would become white in appearance. This is exactly what happens to the filament of a light bulb.

(a) Would an object at a temperature of 3500°C emit more or less radiation than it would at 3000°C? _____

(b) How does the frequency of light emitted depend upon the temperature of the object? _____

Answers: (a) more; (b) As the temperature rises, the maximum frequency becomes greater.

2   Radiation from a solid is referred to (in the ideal case) as blackbody radiation because, it turns out, a perfectly black object is the best radiator of electromagnetic energy. How well did electromagnetic theory explain the radiation curves from blackbodies? This theory predicted that when an object gets hotter, it emits more radiation. And this is indeed the case, as shown in the previous figure, for the curves for higher temperatures rise higher and have more area under them. (The area under the curve shows the total amount of radiation of all frequencies emitted.)

In addition, electromagnetic theory predicted that as an object is heated, the electrical charges within atoms will vibrate faster. The theory linked the frequency of this vibration with the frequency of light emitted by the atom. Thus, the fact that as an object is heated to higher temperatures it emits radiation of higher frequencies also seems to fit with the theory of James Clerk Maxwell (Chapter 18, frames 7, 8).

When physicists tried to derive equations from Maxwell's electromagnetic theory to fit the radiation graphs, however, the equations did not fit, especially at high frequencies. Thus, although qualitatively the theory fit the observations, the equations could not be verified quantitatively. The equations of Maxwell, which described how electromagnetic waves are emitted and travel through space, just could not be fit to the situation of blackbody radiation.

(a) Does the fact that an object heated to a higher temperature emits higher frequencies of radiation support or deny Maxwell's electromagnetic theory?

_____

(b) Do the measurements of frequencies of radiation emitted correspond with those predicted by Maxwell's equations? _____

(c) Does Maxwell's theory depend on a wave or particle nature of light?

_____

Answers: (a) support (at least qualitatively); (b) no; (c) wave

3 PLANCK'S QUANTUM HYPOTHESIS

In 1900, Max Planck, a physicist at the University of Berlin, presented an explanation for the behavior of blackbodies. We saw earlier that as the frequency of vibration of the electric charges within atoms increased, the frequency of radiation emitted was expected to increase. It was expected that any frequency of vibration was possible, just as you can swing a stick above your head at any frequency you want (up to the limit of your physical abilities, anyway). Planck, however, proposed that the electrons in atoms are limited to certain specific energies of vibration. It is as if you could swing the stick at frequencies of 3 swings per second and 4.5 swings per second, but were unable to swing it at any frequency in between. Planck stated that the atoms could exist only in these specific energy states, and when they absorb and release energy they change from one energy state to another. The energy absorbed is thus absorbed in little chunks, or **quanta**. Likewise, he said that when energy is emitted, it is emitted only in quanta. In addition, the energy contained in each quantum is governed by the frequency of the radiation. The formula for the relationship is:

$$E = h \cdot f$$

where E is the energy of a quantum of radiation of frequency f that is absorbed as the blackbody radiates away energy. The symbol h is called Planck's constant and is a number we use as a multiplier of frequency to calculate the amount of energy in a quantum.

The radiation curves shown in the figure of frame 1 can be derived mathematically using Planck's quantum hypothesis. Finally, therefore, blackbody radiation could be explained—by using Planck's ideas.

(a) According to Planck's theory, can an atom gain or lose energy in any amount?

_____

(b) Write Planck's equation. _____

(c) In Planck's formula, what is f the frequency of? _____

_____

Answers: (a) no (only in discrete chunks); (b) $E = h \cdot f$; (c) the quantum of radiation absorbed or emitted by an object

4 EINSTEIN ON THE PHOTOELECTRIC EFFECT

Physicists recognized in 1900 that Planck's hypothesis did indeed explain blackbody radiation, but it was not quickly accepted as being the true nature of things. The quantum idea did not receive much attention until 1905, when Albert Einstein accepted the hypothesis and took it one step further. Planck held that radiation was absorbed and emitted in quanta, but he clung to the idea that radiation itself was continuous. Thus, he pictured Maxwell's electromagnetic waves traveling through space as before, but being emitted and absorbed in small quanta.

Einstein proposed that radiation actually travels through space as quanta of energy. This was another blow to the electromagnetic theory. One now had to somehow think of little chunks of energy (instead of smooth, steady waves) traveling along. Einstein called the chunks "**photons**," and we will use "photon" interchangeably with "quantum" in the rest of this book. Acceptance of the quantum theory does not cause us to abandon the wave theory, however. In fact, the amount of energy in the photons is determined by the frequency of the radiation! As we saw above in Planck's equation, the energy is still directly related to the radiation frequency.

Early in the history of our understanding of nature we wondered whether or not matter was of such a nature that it could be divided into smaller and smaller bits forever without coming to some smallest chunk. We know today that it cannot. There is a smallest chunk into which any substance, at least in principle, can be divided. It is the atom or molecule. Energy, however, had always been considered to be continuous, to occur in any amount, with no limit to how small an amount of energy could be exchanged between two objects. Einstein's idea put energy into the same category as matter, for he stated that energy was also "chunky."

How did Einstein's idea of how energy traveled differ from Planck's?

_____

_____

Answer: Einstein thought energy traveled in photons, while Planck thought it traveled in smooth waves.

5 Four observations about the photoelectric effect were presented in frame 10 of the last chapter. We will review them here from the standpoint of the quantum theory. In 1905, Einstein used Planck's ideas to explain the photoelectric effect. According to Einstein, an electron is ejected from an atom (in the photoelectric effect) when it absorbs a photon of sufficient energy to allow it to free itself from the atom. Instead of gradually gaining energy from electromagnetic waves, the energy transfer is done all at once in quantum jumps. The four observations:

1. Electrons are emitted immediately when light strikes the photoelectric surface.

According to quantum theory, since electrons gain their energy in single steps rather than gradually gaining energy from a wave, this observation holds true. As soon as a photon of sufficient energy strikes an electron, the electron jumps from the metal.

2. High-frequency light causes electrons to be emitted, while low-frequency light does not.

   According to quantum theory, photons of high-frequency light have more energy than photons of low-frequency light ($E = h \cdot f$). Since some minimum energy is needed to free electrons from atoms, apparently photons of low frequency light do not have sufficient energy.

3. The energy of emitted electrons depends upon the frequency of the light that strikes the metal.

   Again, this is easily explained. Light of high frequency has photons of great energy, so when one of these photons is absorbed by an electron, the energy left over above what is needed to free the electron goes toward giving the electron extra speed.

4. More electrons are emitted for bright light than for dim light.

   According to quantum theory, brightness of light depends upon the number of photons in the beam rather than the amplitude of the wave. Therefore, when the light is brighter, more photons cause more electrons to be emitted.

Thus, the photoelectric effect, which was one of the major obstacles to full acceptance of the electromagnetic theory of light, was fully explained by the quantum theory. How does the quantum theory explain each of the following?

(a) Ultraviolet light causes electrons to be emitted while infrared light does not.

_____

(b) A bright blue light causes more electrons to be emitted than a dim one.

_____

Answers: (a) Photons of ultraviolet light have more energy, since ultraviolet light is of a higher frequency than infrared.   (b) Bright light has more photons than dim light, and each photon causes an electron to be emitted.

6 THE BOHR MODEL OF THE ATOM

The figure on the next page illustrates the Bohr model of the atom. In the Bohr model, the atom consists of negatively charged electrons revolving around the central nucleus. Most of the mass of the atom by far is contained in the nucleus, the electrons accounting for less than $\frac{1}{4000}$ of the mass of the atom.

Bohr realized, as did the science community around 1900, that the electromagnetic theory, when applied to the nuclear atom, predicted that electrons should continually be emitting radiation because their orbiting motion is essentially a vibration, and a vibrating electric charge should emit electromagnetic

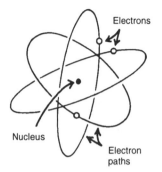

radiation. (This idea was at the heart of the electromagnetic theory and had been verified time and time again. Radio transmission depends upon this idea.) If the atom emitted radiation as predicted, it would gradually lose energy so that the electron would gradually fall in toward the nucleus. The atom would die! Obviously this does not happen.

To solve the problem, Bohr made two assumptions about the atom, and in the process, he explained how photons of light come into being. The two assumptions: (1) electrons that revolve around the nucleus move in specific given orbits, and (2) no radiation is emitted when they are in these orbits. Since each orbit corresponds to a certain energy, the energy levels of an atom are not continuous. When an electron jumps from one allowed orbit to another, the atom absorbs or emits radiation.

(a) For an electron to move to an orbit of greater energy, must it lose or gain energy? _____

(b) When it falls to an orbit of lower energy, does it gain energy or emit it as radiation? _____

Answers: (a) gain; (b) emit

7 | EMISSION OF LIGHT

Part A of the next figure represents an atom with four energy levels for its electrons. (In reality, an atom has many more allowed orbits than this.) The figure shows one electron in the outer orbit. No electrons are in the inner orbits. An atom with an electron in anything but its lowest possible orbit has more than a minimum amount of energy and is said to be **excited**. This is not a stable condition for an atom to be in, and the atom will release its extra energy by means of the electron jumping—or rather falling—down to a lower level.

Part B shows the possible falls that the electron might make. It can fall to levels 3, 2, or 1. If it first falls to 3, it can then fall to 2 or 1. Once in level 2, it will then fall on down to 1. Each time the electron falls from an outer orbit to an inner

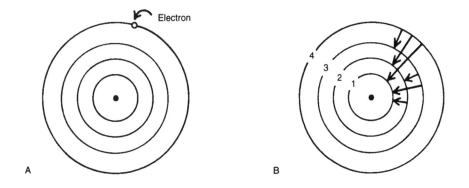

one, it emits an amount of energy that corresponds to the difference in energy of the levels. This energy is released as a quantum (or photon) of radiation.

(a) Examine part B of the figure. What possible electron-fall would you expect to emit a photon of greatest energy? _____

(b) What is the maximum number of photons that this one excited atom could emit? _____

(c) Where will the electron be when the atom is no longer excited? _____

_____

Answers: (a) 4 to 1;   (b) 3 (4 to 3, 3 to 2, 2 to 1);   (c) in the inner orbit

8

The figure in this frame is a representation of the energy levels of a hydrogen atom. Notice that levels of energy rather than circular orbits are shown. The units of energy filled in here are electron volts (eV); don't worry about what the unit means just now. In addition, the energy is negative to conform to standard practice in physics. Suppose the electron is in the –0.85-eV level and falls to the –1.51-eV level. The electron loses energy since the level to which it is falling is lower than its original level. The amount of energy lost is the difference between energy levels: $-0.85 - (-1.51) = 0.66$ eV. Thus, the photon emitted has an energy of 0.66 eV.

(a) Suppose the electron falls from the –0.85-eV level to the –3.40-eV level. How much energy will the emitted photon have? _____

(b) If it falls from the –0.85-eV level to the lowest level, how much energy will the photon have? _____

(c) Suppose that after the electron falls from –0.85 to the –1.51-eV level, it then falls to –13.6 eV. How much energy will the emitted photon have?

_____

Answers: (a) 2.55 eV;   (b) 12.75 eV;   (c) 12.09 eV

9

The energy levels in the figure of the previous frame are, in fact, the energy levels of hydrogen atoms. Thus, when hydrogen atoms emit radiation, the radiation is always of certain characteristic frequencies. Other elements have different sets of energy levels. No two elements have the same set of energy levels. Thus, no two elements emit photons of the same set of energies.

Now recall the tie-in of the quantum theory with the electromagnetic theory. If the energy of two photons is different, the two photons differ in frequency. If we limit ourselves to considering visible light, we see that light of different frequencies appears as different colors. Thus, we find that each element in nature, if heated up to the gaseous state so that it emits light, will emit its own characteristic frequencies—its own characteristic colors.

(a) Could elements be uniquely identified by the color of light their gases emit?

_____

(b) Why or why not? _____

_____

Answers: (a) yes;   (b) Each element has a different set of energy levels, so a different set of frequencies would be emitted.

10

Now suppose we take the light emitted by heated hydrogen gas and pass it through a prism so that it breaks down into its colors. Only certain frequencies of light are present in light emitted by hydrogen atoms, and therefore only certain colors result. The spectrum of light produced from hydrogen atoms looks somewhat like that of the figure here. Note that some of the "colors" of light are

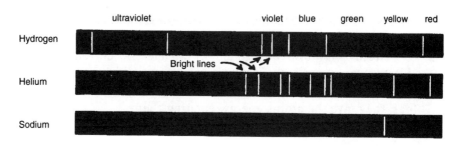

in the ultraviolet region. Although these frequencies are invisible to the eye, they can be recorded on film.

If a different chemical element is used as the source of light, a different spectrum is produced, because each element has its own set of energy levels for its electrons and thus its own characteristic frequencies of emitted light. The characteristic spectra of helium and sodium gas are also shown in the figure. Thus, if upon heating a gas you see the particular combination of frequencies that corresponds to that of helium, as in the figure, you may be sure that the light was emitted by the element helium.

The type of spectrum discussed above is, for obvious reasons, called an **emission spectrum** or a **bright line spectrum**. When light from a glowing gas is separated into its colors (by using a prism, for example), one can determine what elements are present in the gas. This analysis of the spectrum of various materials is done with an instrument called a **spectroscope**, and the technique is called **spectroscopic analysis**.

(a) Refer to the figure. What element shown has the most ultraviolet light in its spectrum? _____

(b) What color light does sodium emit? _____

(c) Why might you want to do a spectroscopic analysis? _____

_____

Answers: (a) hydrogen;   (b) yellow;   (c) to determine what element or elements make up a substance

11 EXCITATION OF ATOMS

We saw how excited atoms lose their extra energy. But how do they get excited in the first place? The simplest way is by heating. When a gas is heated, the atoms move more and more rapidly and collisions become more violent. When collisions become violent enough that there is sufficient energy available to cause an electron to jump to a higher level, it does.

The heating may also be caused by a chemical reaction among the atoms. This is what is happening in a flame. The heat produced by the burning is causing the atoms of the gas to become excited. After excitation, the electrons quickly fall back to an inner orbit and give off their extra energy as light (in the form of photons).

Atoms may also be excited by an electric current passing through a gas. Electric current in a gas is simply electrons and positively charged atoms rushing through the gas. When these fast-moving particles collide with the atoms of the gas, they give the atoms energy, exciting them.

Name three ways of causing atoms to absorb energy and become excited.

Answer: heating, chemical reaction, electric current

12   Let us look at what happens when atoms are excited by fast-moving electrons. The figure here is the energy level diagram for an imaginary atom with an electron in its lowest possible level—called its "ground" state. The figure shows that there are two allowed levels of energy above the ground state.

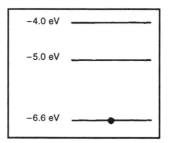

(a) In order to jump from the ground state to the next highest level, how much energy must the electron gain? _____

(b) To jump from that position to the second excited state, how much energy must it gain? _____

(c) How much energy will be released when it finally returns to the stable state?

_____

Answers: (a) 1.6 eV;   (b) 1.0 eV;   (c) 2.6 eV

13   Refer to the figure in the previous frame. Suppose an electron (part of an electric current) is moving with an energy of 1.4 eV and strikes the ground-state electron. Since the approaching electron does not have enough energy to knock the orbiting electron all the way up to its first excited level, it cannot move it at all, and it will therefore not affect it. But suppose the approaching electron has 1.8 eV of energy. In this case, it will give up 1.6 eV of energy upon striking the orbiting electron, causing this electron to jump up to the next level. The striking electron is left with 0.2 eV of energy, and moves on at a reduced speed (corresponding to the 0.2 eV of energy). If the incoming electron has an energy of 2.6 eV or higher, it can cause the orbiting electron to jump up to its second excited level.

Once the orbiting electron has been given energy and is in a high-energy state, it can then fall back down to its original state. In so doing it will release a quantum of energy, as we have seen.

Suppose our stable atom is struck by one electron after another. (Though this is unrealistic, it will emphasize a point.) The approaching electrons have the energies below. What will happen in each case? (Consider this to be a sequence of events.)

(a) 1.5 eV _____

(b) 2.0 eV _____

(c) 0.5 eV _____

(d) 1.0 eV _____

Answers: (a) no effect; (b) The electron jumps to –5.0-eV level, and the striking electron continues with 0.4 eV of energy (moving slower). (c) no effect; (d) The electron jumps to third level, and the striking electron stops.

14 Another way to excite the atoms of a gas is by allowing them to absorb photons of light. This will happen if the photon has just the right amount of energy to cause the electron to make the jump to a higher level. If the photon has too little energy, the electron cannot make the jump. But if the photon has too much energy for a particular jump, the electron will not make that jump either!

Suppose white light (which consists of all frequencies of light) passes through a gas of atomic hydrogen. Any gas, unless it is at a very high temperature, has most of its atoms in the lowest possible energy state, the "ground" state. In the figure of frame 8 we saw that the hydrogen electron has a ground-state energy of –13.6 eV.

(a) What is the least amount of energy that this electron can absorb and jump to a higher orbit? _____

(b) What are three other photon energies that could cause ground-state hydrogen atoms to become excited? _____

Answers: (a) 10.2 eV (13.6 – 3.4); (b) 12.09, 12.75, and 13.15 eV (Actually, all four could happen simultaneously in separate atoms.)

15 Now, remember that we said that we were shining white light through the gas. The gas, however, is absorbing those frequencies that correspond to jumps that its electrons can make. Thus, the light that passes through the gas will lack the frequencies that correspond to the photons absorbed. The spectrum of light getting through is similar to the figure in this frame. This figure shows a continuous

spectrum containing all frequencies except those absorbed. (The shaded areas indicate light of the colors stated above them.) Such a spectrum is called an **absorption,** or **dark line, spectrum.** (Recall that a heated gas produces a bright line spectrum.)

Since the frequencies of absorption are fixed by the possible electron jumps, the dark lines appear at particular characteristic positions in the spectrum, the same positions where the bright lines appeared in the emission spectrum of that particular chemical element. Thus, an absorption spectrum can also be used to identify elements. The light coming from stars passes through the nonglowing gases near the star as well as through the earth's atmosphere before reaching us. Therefore, it appears as an absorption spectrum. If we eliminate from consideration those absorbed frequencies that are due to the earth's atmosphere, we can examine the spectrum of light from a star and determine what chemical elements are in the star's atmosphere. This is a standard spectroscopic procedure in astronomy.

(a) How does an emission spectrum compare to an absorption spectrum for the same element? _____

_____

(b) Why do the dark lines of the absorption spectrum appear at the same place where the emission spectrum has bright lines? _____

_____

Answers: (a) The emission is bright for certain frequencies while the absorption is dark for the same frequencies;   (b) They result from electrons jumping between the same levels. (In one case the electron jumps "up" and in the other case it falls "down.")

16 TYPES OF LAMPS

A regular incandescent light bulb, which is made with a filament of tungsten, emits white light rather than frequencies characteristic of the element tungsten. This occurs because the tungsten filament is a solid rather than a gas. In a solid, the individual atoms are not independent of one another, but instead are linked closely together. Thus, the electrons of an atom are not influenced exclusively by their "home" atom. In fact, electrons jump from one atom to another in metals and have no "home." As a result, a photon of essentially any energy can be emitted in one electron jump or another. The outcome of this is a **continuous spectrum,** containing all the colors of the rainbow.

(a) In a solid are there more or fewer possible energy jumps? _____

(b) What color light emission results when electrons are not restricted as to energy levels? _____

Answers: (a) more; (b) white

17

Some street lights have a more bluish tint than others. These lamps are most likely mercury vapor lamps. As their name suggests, they contain a vapor of the element mercury. When the mercury is heated by an electric current, it emits light. The mercury vapor lamp does not show a continuous spectrum, but instead shows an emission (bright line) spectrum. And the frequencies in this spectrum are characteristic of the element mercury.

In some parts of the country you see bright yellow street lamps. These are probably sodium vapor lamps, containing the vapor of the element sodium. This element, when heated to a vapor form, emits only yellow light. (See the figure in frame 10 for the sodium spectrum.)

(a) Why does a gas vapor lamp not emit white light as does the incandescent lamp? _____

_____

(b) Our eyes are more sensitive to yellow light than any other color. What advantage does this suggest for sodium vapor lamps? _____

_____

Answers: (a) Gas vapor lamps produce emission spectra of certain frequencies, while solids in incandescent lamps don't. (b) Since sodium vapor produces only yellow light, they appear brighter to us than other lamps of the same wattage. (In fact, they are 6 to 8 times more efficient than incandescent lamps.)

18

If you pass the light from a fluorescent tube through a prism and observe its spectrum, you will see a combination of a bright line spectrum and a continuous spectrum. The bright line spectrum is easy to explain—the tube contains gas, primarily mercury. To see why the continuous spectrum also shows up, we must consider the phenomenon of fluorescence.

Most solids, when light shines on them, will either reflect the light or absorb it and then emit the same frequency they absorbed. Certain substances, called fluorescent, have the property of being able to absorb high-frequency light—high-energy photons—and then emit light of a lower frequency. Such a fluorescent substance is used to coat the inside of the tube of the fluorescent lamp. Some of the frequencies emitted by mercury gas are in the ultraviolet region of the spectrum. Suppose the atom of the figure in this frame absorbs a photon of ultraviolet light, causing the electron to jump up to the third excited level. When the electron falls

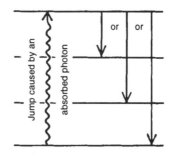

back down to a lower state, however, it can fall to any of the three states below it. And since the energy it loses if it falls to the first or second excited state is less than if it falls all the way to the ground state, the frequency associated with this fall will be less than the frequency of the ultraviolet light that started the whole process. The result is that the ultraviolet radiation emitted by the mercury causes the tube's coating to fluoresce in the visible regions of the spectrum.

(a) Would you expect the solid coating to emit a continuous or bright line spectrum? _____

(b) What causes the bright line spectrum in a fluorescent lamp? _____

_____

(c) Fluorescent lamps do not get as hot as incandescent lamps. Would you expect a 40-watt fluorescent lamp to emit more or less light than a 40-watt incandescent lamp? (The wattage of a bulb tells us the electrical energy used each second.) _____

Answers: (a) continuous (see frame 16); (b) the mercury vapor; (c) more (because less electrical energy is transferred to heat energy)

---

19   The phenomenon of fluorescence is also responsible for the behavior of "black light" posters. "Black light" is a common term for ultraviolet light. When such a poster is exposed to ultraviolet radiation, it absorbs the ultraviolet and emits visible light. Since our eyes are not sensitive to ultraviolet, "black lights" themselves do not appear bright to us. But some of the energy they emit is absorbed and then emitted as visible light by the fluorescent dyes on the poster, so the poster itself glows.

(a) A "black light" appears violet in color. What do you think causes this?

_____

(b) How is a "black light" poster similar to a fluorescent lamp? _____

_____

Answers: (a) visible violet light from the high end of the visible spectrum (not ultraviolet waves); (b) Both absorb high-energy photons and emit lower energy photons.

20 PHOSPHORESCENCE

Some materials have the property of absorbing a photon of light but not re-emitting the energy immediately, as if the electron jumped to an excited state and got stuck there for a while. Suppose you have a number of atoms of such a material, and you expose them to light. Many of them will absorb photons and become excited, and gradually, atom by atom, they will emit visible light. You then turn off the source of light. The material will continue to glow as some of the atoms are still in the excited state and one by one emit their extra energy in the form of visible light. Such substances are called **phosphorescent,** and are used for such purposes as coatings for the hands and numbers on clocks.

Which of the common items below are phosphorescent?

_____ (v) lighted dials on electric alarm clocks

_____ (w) glow-in-the dark Christmas tree ornaments

_____ (x) black light posters

_____ (y) bright paint on a new car

_____ (z) paint on the hands of a wristwatch

Answer: (w) and (z)

21 THE LASER

Normally, when an electron is excited to an outer orbit, it falls back to an inner orbit, causing a photon of light to be emitted. Although the time delay between excitation and de-excitation is extremely short (except for phosphorescent substances), it does exist. What determines how long the time will be for any individual atom? Apparently, the fall back to an inner orbit is a chance occurrence and happens spontaneously. But there is a method by which the excited atom can be caused to de-excite and emit its photon.

Part A of the figure in this frame shows an excited atom about to be struck by a photon of exactly the same frequency as the photon that will be released when the electron falls inward. Such a photon will "stimulate" the atom to de-excite and emit energy. Amazingly, the emitted photon will come off in the same direction as the one that caused the emission and in phase with that one. Part B shows what this means. The two photons are traveling away "in step." This is the secret of the laser, the word being an acronym for Light Amplification by Stimulated Emission of Radiation.

Part C of the figure shows how we might think of the waves of light from a regular incandescent lamp. Since each photon of light is emitted spontaneously, they are not in step with one another, but are all jumbled. In addition, they are of

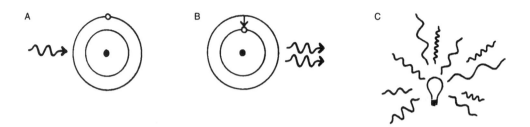

various wavelengths. They are said to be "incoherent." Laser light, on the other hand, is in phase (in step), and is of one single wavelength. It is "coherent."

(a) Which involves spontaneous emission of energy? _____

(b) What is the source of the word *laser*? _____

Answers: (a) incandescent; (b) light amplification by stimulated emission of radiation

22 Although there are differences from one type of laser to another, all have one basic feature in common: the need to have many excited atoms available to be stimulated to emit light. The laser itself may be either solid or gas. It consists of a tube containing atoms that have one excited state in which electrons tend to stick—they do not readily fall to an inner orbit. On both ends of the tube are mirrors.

To operate the laser, the atoms are excited by either a flash of light or an electric current. Electrons become fixed in the orbit referred to above until finally (actually a very short time by human standards) one of them de-excites and emits a photon. If this photon is going in any other direction except down the length of the tube, it will escape from the tube and not cause laser action. Eventually, however, a photon will be emitted down the tube. When it passes another excited atom, it will stimulate that atom to emit a photon and the two photons will move off in phase. Then each of them will encounter another atom. Result: four photons in phase, then eight, 16, 32, and so on. When they reach the end of the tube, they reflect back again and the laser action continues to grow. This would be of no value if it were confined to the laser tube, but one of the mirrors is made so that a very small fraction of the light hitting it escapes through it. This is the laser beam that today is used for everything from surveying to eye surgery.

The advantage of laser light, then, is that it is intense because of its waves being in phase. (If you studied Chapter 12, you realize that out-of-phase waves cancel one another.) In addition, the light remains concentrated—it does not spread out much as it travels along.

(a) Will just any photon emitted in the laser tube start the laser action? _____

(b) Where is the light emitted from the laser tube? _____

_____

(c) What is special about photons of laser light? _____

_____

Answers: (a) no (it must be moving down the length of the tube);   (b) through one of the mirrors on the ends;   (c) They are in phase and have the same wavelength.

SELF-TEST

1. How does the frequency of the radiation emitted depend upon the temperature of the source? _____

_____

2. What did electromagnetic theory predict concerning the relationship between the temperature of an object and the frequency of radiation emitted? How well did the theory fit the experimental data? _____

_____

3. Who first suggested the quantum idea that explained blackbody radiation?

_____

4. Explain Planck's quantum hypothesis. _____

_____

5. How did Einstein expand Planck's original quantum idea? _____

_____

6. How does quantum theory explain the observation that high-frequency light is able to cause electrons to jump free of some metals while low-frequency light cannot do so? _____

_____

7. According to quantum theory, upon what does the brightness of light depend? _____

8. Explain, according to Bohr's model of the atom, how light is emitted by the atom. _____

_____

9. What type spectrum is produced by a heated gas? _____

_____

10. What practical application is made of the fact that the spectrum of one element is different from the spectrum of another? _____

_____

11. Refer to the figure of frame 12. Suppose an electron with an energy of 2.8 eV strikes the orbiting electron. How much energy might this electron lose in the collision? _____

12. What is it about a fluorescent lamp that causes it to show a continuous spectrum (along with the bright line spectrum)? _____

_____

13. What is "black light," and how does it make fluorescent paint glow? _____

_____

14. How does phosphorescence differ from fluorescence in the manner in which the atom emits light? _____

_____

15. What causes laser light to be more intense than regular light? _____

_____

ANSWERS

1. As the temperature of the source becomes greater, the frequency of radiation is increased. (frame 1)
2. Electromagnetic theory predicted that frequency would increase as temperature increases, but the prediction could not be made to fit the data in a quantitative way. (frame 2)
3. Max Planck (frame 3)
4. Planck hypothesized that atoms lose and gain energy only in definite, fixed jumps, which he called quanta. (frame 3)
5. Einstein proposed that light itself traveled as photons (quanta) of energy. (frame 4)
6. Low-frequency light is made up of photons of low energy. Their energy may not be enough to release the electrons, but light of higher frequency has photons of higher energy, and these may be energetic enough to knock electrons from atoms. (frame 5)
7. The brightness of light depends upon the number of photons in the beam. (frame 5)
8. Light is emitted as photons when an electron falls from an orbit of greater energy to an orbit of lesser energy. The energy of the photon is equal to the difference in energy between the two orbits. (frames 6, 7)
9. An emission (or bright line) spectrum is produced by a heated gas. (frame 10)
10. This fact can be used to identify the elements that compose a substance. (frame 11)
11. either 1.6 eV or 2.6 eV (frames 12, 13)
12. The coating inside the tube is a solid, and it emits photons of many energies when it fluoresces. (frame 18)
13. "Black light" is the name sometimes given to ultraviolet radiation. The atoms of the fluorescent paint absorb photons of ultraviolet radiation and re-emit photons of lower energy—visible light. (frame 19)
14. In the case of fluorescence, the atom re-emits its energy (almost) immediately. In phosphorescence, the atom may wait quite a while before its electron returns to a lower energy level and emits the energy as light. (frames 19, 20)
15. The fact that it is coherent (its waves are all in phase). (frames 21, 22)

20 Reflection, Refraction, and Dispersion

No prerequisites; Chapters 10 and 11 suggested
The phenomena that form the subject of this chapter—reflection, refraction, and dispersion of light—are among the most common of our everyday experience. The first two are so common, in fact, that we are seldom aware of their existence. The third is what gives us the beautiful colors of the rainbow.

OBJECTIVES

After completing this chapter, you will be able to

- compare the angles of incident and reflected light;
- differentiate between specular and diffuse reflection;
- determine relative size and position of the object and image in a plane mirror;
- differentiate between reflection and refraction;
- explain how the speed of light affects refraction;
- compare the wavelengths of light in two materials where the speed of light differs;
- specify the meaning of "total internal reflection" and state the conditions necessary for its occurrence;
- explain the effect of refraction on our perception of sunset;
- explain the cause of mirages in terms of the phenomena presented in this chapter;
- specify the relationship of dispersion to refraction;
- state the cause of rainbows;

- determine the angle of refraction, given the angle of incidence of light and the speed of light in each material (optional);

- determine the angle of refraction, given the angle of incidence of light and the index of refraction of each material (optional);

- determine the critical angle for total internal reflection between two substances, given the index of refraction of each (optional);

- calculate the speed of light in a substance given the index of refraction of the substance (optional).

1 REFLECTION

If you shine a flashlight into a mirror in a dark room with a lot of dust or smoke in the air, you will be able to see the beam approach and reflect from the mirror. It will appear somewhat as in the figure here. The dashed line is drawn perpen-

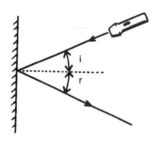

dicular to the mirror at the point where the beam strikes it. The line is said to be "normal" to the surface and is often called "the normal." Notice that the angle that the incoming (or "incident") beam makes with the normal is equal to the angle made by the reflected beam. This relationship between the angle of the incident beam (angle i) and the angle of the reflected beam (angle r) is called the law of reflection and is stated in formula form as:

$$i = r$$

(a) According to the law of reflection, if a beam of light strikes a mirror at an angle of 35° with the normal, at what angle will the reflected beam leave the mirror? _____

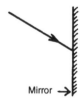

Mirror →

(b) The figure here represents a beam of light striking a mirror. Sketch in the normal and the reflected beam.

Answers: (a) 35° with the normal;   (b) The two angles marked in the figure here are equal.

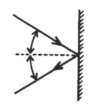

2

When light is reflected from a plane (flat) mirror, the beam is reflected evenly as **specular reflection**. When light strikes a nonshiny surface such as the paper of this page, however, **diffuse reflection** occurs. The figure in this frame represents the surface of this page highly magnified. The light that hits the surface strikes tiny parts of the surface that are at different angles to the incoming rays. Since there is no regularity in the surface, there is no regularity in the reflected light—it bounces off in all directions. Diffuse reflection allows you to see this page (and most other objects) no matter at what angle they are with respect to you and to the incoming light.

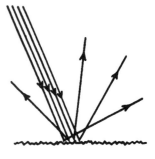

(a) What type of reflection allows you to see this book as you read? _____

(b) What type of reflection allows you to see yourself in a mirror? _____

(c) What is the formula for the law of reflection? _____

(d) State the law of reflection in words. _____

_____

Answers: (a) diffuse; (b) specular; (c) angle i = angle r, or $i = r$; (d) The incident angle (with the normal) is equal to the reflected angle (again measured with respect to the normal).

3

The figure in this frame shows two rays of light coming from the head of a nail and bouncing from a plane mirror. Light is coming from the nail in all directions, of course, but some rays happen to strike the mirror and bounce off into the two eyes at left.

To the two eyes, the light entering them seems to have come from a single point behind the mirror. If the eyes are moved so that two more rays of light coming from the nail bounce into the eyes, these two rays will still seem to have come from exactly the same point as the two rays in the figure. In fact, every ray of light that leaves that point on the nail and strikes the mirror will rebound in a direction as if it came from the corresponding point behind the mirror.

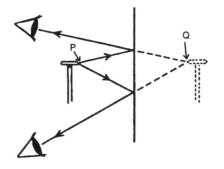

(a) What type of reflection are we discussing? _____

(b) How does the distance from the mirror to point P compare to the distance from the mirror to point Q? _____

Answers: (a) specular; (b) The distances are equal.

4 We can analyze the light that comes from any point on the nail in the figure in the previous frame. All of the reflected light will seem to have come from a corresponding point behind the mirror. In this way an **image** of the nail is formed by the mirror. In the figure, the distance from the nail to the mirror is the same as the distance from the image to the mirror and the image is the same size as the original nail.

These observations can be generalized to a law: The image of an object formed by a plane mirror is the same size as the object and is located behind the mirror at a distance equal to the object's distance in front of the mirror.

(a) What type of mirror forms an image the same distance behind it as the object is in front of it? _____

(b) The object shown below is near the mirror. Sketch the image.

Answers: (a) piane; (b) The image you drew should be the same size as the object (the heart) and it should be an equal distance from the mirror, on the opposite side.

5 REFRACTION

Figure Q (on the facing page) shows the beam from a flashlight being shined into water. Again the normal is drawn at the point where the beam enters the water. Compare the angles made by the incident light and the light after being bent—refracted—at the surface. Notice that the angle outside the water is greater.

Figure R shows a ray of light entering a flat plate of glass from the upper left. Notice that upon entering the glass it behaves just as did the beam entering water in the last figure. It bends so that its angle with the normal is less inside the glass than outside. Now let us focus on what happens when the light emerges from the glass. Again, a normal is drawn to the point where the light crosses the surface.

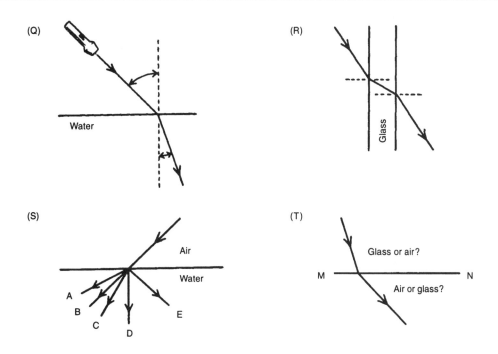

(a) Compare the angles involved as the ray comes out of the glass. Which angle is largest? _____

(b) In figure S, a ray of light is shown about to enter water. Refer back to Figure Q. In which of the directions—A, B, C, D, or E—will the ray go after it crosses the surface? _____

(c) Figure T shows the path taken by a ray of light as it crosses the boundary between glass and water. (Line MN represents the surface.) Is the glass above or below the line MN? _____

(d) Generalize from the three examples: When light crosses a boundary between air and water or air and glass, it bends so that the angle (with the normal) in the air is (greater/less) _____ than the angle in the other material.

Answers: (a) the angle outside the glass; (b) C; (c) above; (d) greater

6

In order to see what causes light to bend when it crosses a surface, we will consider light as a wave rather than considering single rays of light. The rays are simply infinitesimally narrow beams of light. Their use is a convenient way to emphasize the direction light is traveling. The figure on the next page shows wavefronts (concentric circles) and rays coming from a point source of light. The rays are always perpendicular to the wavefronts.

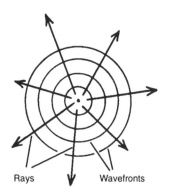

Rays   Wavefronts

(a) Did we consider rays or waves to study reflection of light? _____

(b) Does a ray of light travel in the same direction as the wave travels? _____

(c) How does reflection differ from refraction? _____

_____

_____

Answers: (a) rays; (b) yes (they are just two different ways to consider the light); (c) Refraction involves light going from one material into another. Reflection involves bouncing from the surface of another material.

7

The figure in this frame represents straight wavefronts moving in air from upper left to lower right and striking a surface (MN) of glass. The speed of light in glass is less than that in air, so the waves will slow down upon crossing the boundary. Look at what happens to wave CD. Although when it was at position AB, it was a straight wavefront, end C has already entered the glass, where it travels slower; this results in its covering less distance than end D. And the result of this is that the wavefront is no longer straight. The portion from C to the surface is now closer to parallel to the surface. Finally, end D will enter the glass and the wavefront will be in a position such as EF. It is once again straight. But since the wave travels in a direction perpendicular to the wavefront, it is now heading in a different

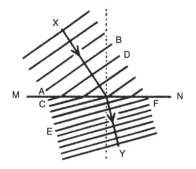

direction than it was before it entered the glass. The direction of travel of the light both before and after entering the glass is shown by the ray that stretches from X to Y.

(a) How does the angle with the normal of the ray compare before and after entering the glass? _____

(b) What causes the light ray to change direction in this example? _____

_____

Answers: (a) larger outside the glass; (b) the change in speed of light

8 Imagine that the waves of the figure in the previous frame are moving in the opposite direction—from lower right (in the glass) to upper left (in air). In this case, side D emerges first from the glass and increases in speed, resulting in an increase in wavelength and a change in direction. Again, the law of refraction stated in an earlier frame is explained by the fact that light travels at different speeds in different materials. We might state the law of refraction qualitatively as follows: When light crosses a boundary between materials in which it travels at different speeds, it bends such that its angle with the normal is less in the material in which it travels more slowly.

(a) Refer to the figure of frame 7 again. Does light travel slower above or below the line MN? _____

(b) Is its angle with the normal less above or below line MN? _____

(c) Light travels more quickly in a vacuum than in clear plastic. If light crosses from a vacuum into plastic, in which medium will it have the greater angle with the normal? _____

Answers: (a) below; (b) below; (c) the vacuum

9 CALCULATING ANGLES OF REFRACTION (OPTIONAL)

Thus far we have considered refraction only qualitatively. The quantitative statement of the law of refraction involves sines of the angles between the light rays and the normal. Use a calculator containing trigonometry functions to answer these questions.

(a) What is the sine of 20° (which is usually written "sin 20°")? _____

(b) What angle has a sine of 0.575? _____

Answers: (a) 0.342; (b) 35° (Actually, it is a little greater than 35°, but we will use the nearest whole degree.)

10   The bending of light as it passes from one material to another is caused by the fact that light has a different speed in the two materials. If we consider light passing from material A into material B, as shown in the figure, the law can be stated as follows:

$$\frac{\sin \theta_A}{\sin \theta_B} = \frac{v_A}{v_B}$$

where θ_A and θ_B are the angles made with the normal by the rays of light in materials A and B, respectively, and v_A and v_B are the speeds of light in materials A and B.

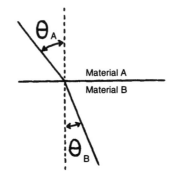

The speed of light in air is about $3 \cdot 10^8$ meters/second* and in water it is $2.25 \cdot 10^8$ m/s. Suppose a ray of light in air strikes the surface of water at an angle of 30° with the normal. At what angle with the normal does it travel in the water? (Use a calculator.) _____

Answer: 22°. Solution: Since it doesn't matter which is material A and which is B, we'll let water be A and air be B.

$$\frac{\sin \theta_A}{\sin 30°} = \frac{2.25 \cdot 10^8 \text{ m/s}}{3.0 \cdot 10^8 \text{ m/s}}$$

$$\frac{\sin \theta_A}{0.5} = 0.75$$

$$\sin \theta_A = 0.375$$

$$\theta_A = 22°$$

11   The quantitative law of refraction is usually stated differently from the equation in frame 10. To understand the way it is usually stated we must first define the index of refraction.

---

*Appendix I is a discussion of powers-of-ten notation.

The index of refraction, n, of a material is the ratio of the speed of light in a vacuum, c, to the speed of light in that material. Or:

$$n_A = \frac{c}{v_A}$$

In the example of frame 10, the index of refraction of water is $(3 \cdot 10^8)/(2.25 \cdot 10^8)$, or 1.33. From the equation defining the index we can see that

$$v_A = \frac{c}{n_A}$$

Dividing this by a similar expression for the velocity in material B, we get

$$\frac{v_A}{v_B} = \frac{c/n_A}{c/n_B} = \frac{n_B}{n_A}$$

Thus, we can rewrite the law of refraction as

$$\frac{\sin \theta_A}{\sin \theta_B} = \frac{n_B}{n_A}$$

Or, as commonly stated:

$$n_A \cdot \sin \theta_A = n_B \cdot \sin \theta_B$$

Table 20.1 gives the indexes of refraction for a number of common substances.

Table 20.1

Substance	Index of refraction
water	1.33
air	1.0003
crown glass	1.52
flint glass	1.66
diamond	2.42

(a) Refer to the table. What is the speed of light in crown glass? _____

(b) Suppose material B in the figure of frame 10 is diamond and that a ray of light traveling in air strikes the diamond at an angle of 30° with the normal. What is the angle of the refracted ray? _____

(c) Compare the answer of (b) above to the answer to the question in frame 10. In both cases the angle in air was 30°. Why does the ray refract (bend) more in diamond? _____

(d) We have said that the speed of light in air is about the same as in a vacuum. How does Table 20.1 show this? _____

Answers:
(a) $1.97 \cdot 10^8$ m/s. Solution: $v = c/n = (3 \cdot 10^8$ m/s$)/1.52$;
(b) 12°. Solution: $n_A \cdot \sin \theta_A = n_B \cdot \sin \theta_B$

$$1 \cdot \sin 30° = 2.42 \cdot \sin \theta_B$$
$$\frac{0.5}{2.42} = \sin \theta_B$$
$$0.206 = \sin \theta_B$$

(c) The index of refraction is greater for diamond than for water. (Or, the speed of light is less in diamond than in water.)
(d) The index for air is almost equal to 1.

12

In this frame we will work through another problem in detail. Suppose a ray of light is traveling in the water of an aquarium and strikes the glass wall at an angle of 25° (with the normal). We want to know first its angle in the (crown) glass and then in the air. See the figure here.

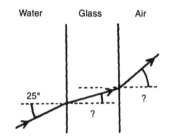

(a) First solve for the angle in the glass, using the equation for the law of refraction. _____

(b) Once you know the angle made with the normal in the glass, you know the angle at which the ray strikes the glass–air surface on the right—the same angle. Now what is the angle in air at the right? _____

Answers:
(a) 22°. Solution: Letting the glass be material A:

$$1.52 \cdot \sin \theta_A = 1.33 \cdot \sin 25°$$
$$\sin \theta_A = \frac{1.33}{1.52} \cdot 0.423$$
$$\sin \theta_A = 0.370$$
$$\theta_A = 21.7°$$

(b) 34°. Solution: Letting A be the air this time:

$$1.0 \cdot \sin \theta_A = 1.52 \cdot \sin 21.7°$$
$$\sin \theta_A = 1.52 \cdot 0.370$$
$$\sin \theta_A = 0.562$$
$$\theta_A = 34.2°$$

13 TOTAL INTERNAL REFLECTION

In the figure below, part A shows a ray of light passing from glass into air. It makes an angle of 30° with the normal in glass and about 50° in air. In figure B, the corresponding angles are 35° and 60°. (If you studied the quantitative refraction section, you can check these out, using an index of refraction of 1.52.) In figure C, the angles are 40° and 78°. In D they are 41° and 86°.

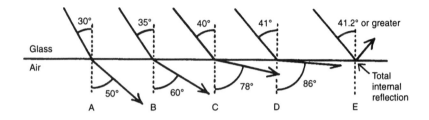

The angle in the air is approaching 90°. Once it reaches this point, no light is refracted, but all of it is reflected back into the material from which it hit the surface (figure E). This phenomenon is called **total internal reflection**. It occurs when two conditions are met:

1. The light is moving toward the surface of a material in which it travels faster; and

2. It hits the surface at an angle (with the normal) greater than some critical angle.

The critical angle for light moving in the glass above toward an air surface is 41.2°.

(a) What happens if light traveling in this glass comes to the air–glass surface at an angle less than 41.2°? _____

(b) What happens if it hits the surface at an angle greater than 41.2°?

_____

(c) What happens if a light beam is moving in air toward a glass surface at an angle of 60°? _____

Answers: (a) It is refracted through the surface. (b) It is totally internally reflected. (c) It is refracted into the glass. (Condition 1 is not met—the light moving toward the surface is in air and is not inside, "internal to," the glass.)

14 Total internal reflection is used in a practical application in optical fibers. The figure here represents a solid cylindrical filament of glass (or plastic). Light has entered the left end, and at each encounter with a surface, its angle with the normal is greater than the critical angle at which it can emerge from the glass (or plastic). This results in the light being carried on down the glass to the other end.

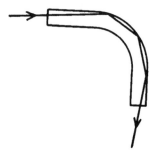

Optical fibers are used routinely today in medicine, communications, and industry. They replace wires to carry telephone and cable television signals. Instead of electricity in wires, light carries the signal in extremely fine optical fibers. The advantages are that there is little loss of power and resultant heating of the optical fibers as there is for wires, and a single light signal is capable of carrying a very great number of communications at the same time (by using different frequencies for different signals).

(a) How does an optical fiber differ physically from a hollow plastic tube?

_____

(b) Why doesn't the light leave the optical fiber? _____

_____

Answers: (a) It is not hollow, but solid.   (b) The light strikes the surface at greater than the critical angle.

15 QUANTITATIVE TREATMENT (OPTIONAL) (Prerequisite: frames 10, 11)

The critical angle for total internal reflection is the angle that would make the light refract at an angle of 90° with the normal. If we let material A be the "internal" material, we can write:

$$n_A \cdot \sin \theta_A = n_B \cdot \sin 90°$$

Since the sine of 90° is 1.00, we can write:

$$\sin \theta_{crit} = \frac{n_B}{n_A}$$

Where the external material is air, we have:

$$\sin \theta_{crit} = \frac{1}{n_A}$$

Consider the critical angle for a water–air boundary.

(a) In which material would the phenomenon of total internal reflection occur? _____ Why? _____

(b) What is the critical angle? _____

Answers:
(a) water; Light travels slower in water than in air.

(b) 49°. Solution: $\sin \theta_{crit} = \dfrac{1}{1.33}$

$\sin \theta_{crit} = 0.75$

16 ATMOSPHERIC REFRACTION: SUNSET

A definite boundary between the two materials is not actually necessary to produce refraction. The phenomenon will also occur if there is a gradual change in a material from one point to another so that light travels at different speeds at the two points. We have been considering the speed of light in air to be the same as the speed of light in a vacuum, but actually light travels slightly slower in air.

In the figure below, light from the setting sun is shown entering the atmosphere and bending down to an observer. Notice that the light that strikes the person seems to be coming to her from higher in the sky than the actual position of the sun. She sees the sun as being higher in the sky than it actually is! In fact, we can see the sun after it has set (if we define "set" based upon a straight line from the sun to us).

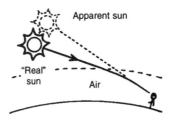

(a) What phenomenon enables us to see the sun after it has set? _____

(b) What causes this phenomenon? _____

_____

Answers: (a) refraction; (b) Light travels slower in air than in a vacuum.

17 The slight difference in speed between a vacuum and air means that atmospheric refraction is not very great until the angle at which the light strikes the atmosphere is great. This is noticeable on a very clear evening just before the sun seems to be setting on a low horizon (the ocean, perhaps). In such an instance, the light from the "bottom" of the sun is bent more than the light from the "top," and the sun appears to be flattened. The next time you see a really clear sunset, look for this. Although refraction by the atmosphere is not normally great, astronomers who are plotting star positions must take it into account, for in precise measurements the refraction is enough to make a difference.

When the setting sun appears to be flattened, are light rays from the top or bottom more strongly refracted? _____

Answer: bottom

18 MIRAGES

Light travels slightly faster in hot air than in cooler air. In the figure below, the air near the hot road is hotter than the air farther above. This results in light from the sky being refracted back up from the road. The ray of light in the figure bends up to the eye of the driver. To the driver, it appears that the road is *reflecting* blue light from the sky, and he is likely to think that the road ahead appears to be wet. Look for this the next time you ride on a hot asphalt road on a still summer day (with no wind to stir the air).

Light from sky

This is the same phenomenon that results in thirsty desert travelers seeing waterholes ahead, only to find them disappearing when they are approached. In this case, imagination also plays a bit role.

What phenomenon causes mirages? _____

Answer: refraction

19 DISPERSION

In what has been said so far, we have ignored the fact that all frequencies of light do not travel at the same speed in all materials. It is true that all frequencies travel at the same speed in a vacuum, but in all other materials the speed of the lower frequencies is greater than the speed of higher frequencies. Since the visible spectrum goes from red at the lowest frequency through orange, yellow, green, blue,

and finally violet at the greatest frequency, we would expect the various colors to be bent at different angles when they are refracted at a boundary. Low frequencies have the greater speed in glass, so we would expect violet to bend most when light enters glass. (Because violet travels more slowly in glass, it slows down more upon entering. It is this slowing down that causes the bending.) And indeed, this separation of colors occurs at every refraction of light. It is not observed as light passes through a window for two reasons. First, the separation by thin windows is very slight. Second, when the light emerges, the violet that was separated from the red of one ray emerges with the red of another ray and no one is the wiser.

The figure at right shows light going through a prism. In this case, the sides of the glass are not parallel, so that the effect produced by the first side is enhanced by the second side. This separation of light into colors by refraction is called **dispersion**.

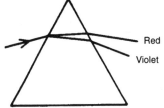

(a) Is dispersion more apparent when light passes through a flat sheet of glass or through a prism? _____

(b) Which color light is refracted most by glass? _____   Why? _____

_____

(c) Is there light between the red and violet that is not indicated in the figure?

_____

Answers: (a) prism;   (b) violet; It travels slower in glass than do the other colors.   (c) yes, the entire spectrum

20 | THE RAINBOW

The figure below shows a ray of white light striking a spherical drop of water at point A. The light is dispersed slightly, with the various colors hitting the other

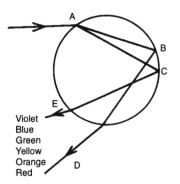

side of the drop between points B and C. At these locations some of the light is refracted out of the drop, but we are interested in the part of the light that is reflected back toward the region from D to E. Here again some light is refracted out of the drop and some is reflected. It is the light that emerges at the surface between D and E that results in rainbows.

The second figure in this frame shows two drops being hit by sunlight coming from the left. Light that reaches the person's eye from the higher drop is red, while light from the lower drop is violet. Raindrops in between these send light of the other colors of the spectrum to the rainbow-watcher. Since the angles necessary to see each of the colors occur not only along a vertical line, but to the sides as well, a curved bow is seen.

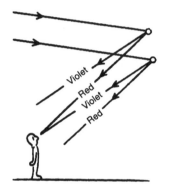

(a) What phenomenon of this chapter causes rainbows? _____

(b) Why must the sun be shining for you to see a rainbow? _____

_____

(c) If you are to see a rainbow at sunrise, what direction must you look? _____

Answers: (a) dispersion; (b) to provide the light (that comes from a single direction); (c) west (Since the sun rises in the east, the rainbow must be opposite to it. The rainbow thus arches across the west.)

SELF-TEST

1. What is the law of reflection? _____

_____

2. What is the difference between diffuse and specular reflection? _____

_____

_____

3. (a) What is the relationship between the size of the object and the image that is formed by a plane mirror? _____

   (b) How does the distance from the object to a plane mirror compare to the distance between the image and the mirror? _____

   _____

4. Suppose you find a figure such as the one here in a textbook. It represents light passing through a boundary between glass and air. The author, however, has forgotten to tell you on which side of the surface is the glass. Is the glass above or below the surface? _____

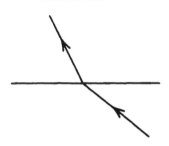

5. Suppose light travels from air into glass. Is its wavelength greater in the air or in the glass? _____

6. What conditions are necessary in order for the phenomenon of total internal reflection to occur? _____

   _____

   _____

7. What phenomenon causes us to be able to see the sun after it has set below the horizon? _____

8. Are mirages caused by reflection or refraction? _____

9. Blue light travels slower than yellow light in glass. Which of the two colors will be bent more by a glass prism? _____

10. Although both reflection and refraction must be used in the explanation of the rainbow, which causes the dispersion? _____

11. *Optional.* Does light travel faster in crown glass or in flint glass? (Use Table 20.1.) _____

12. *Optional.* Suppose light traveling in water strikes the boundary with air at an angle of 25° (with the normal). The index of refraction of water is 1.33. At what angle does it travel in the air? _____

13. *Optional.* Is the critical angle for diamond greater or smaller than that for glass? (Use Table 20.1.) _____

14. *Optional.* What is the critical angle for a diamond–air boundary? (The index of refraction of diamond is 2.42.) _____

ANSWERS

1. The angle of incidence is equal to the angle of reflection. (frame 1)

2. Specular reflection is the regular, orderly reflection such as occurs in the case of mirrors. Diffuse reflection is reflection in all directions, such as occurs from a nonshiny surface. (frame 2)

3. (a) The object and image are the same size.

   (b) The object and image are the same distance from the mirror. (frame 3)

4. above (The ray is closer to the normal above the boundary (not shown), so we know that the light travels slower there.) (frame 5)

5. The wavelength is greater in air because light has a greater speed there. (frame 7)

6. The light must be coming toward the material in which it moves faster, and it must be incident upon the surface at an angle greater than the critical angle. (frame 13)

7. refraction (frame 16)

8. refraction (frame 18)

9. The blue light will bend more. (frame 19)

10. refraction (frame 20)

11. crown glass (Since crown glass has the lesser index of refraction, it follows from the definition of the index that light travels faster in that material.) (frame 11)

12. 34°. Solution: calling air material A:
$$1.000 \cdot \sin \theta_A = 1.33 \cdot \sin 25°$$
$$\sin \theta_A = 1.33 \cdot 0.423$$
$$= 0.563 \text{ (frames 11, 12)}$$

13. smaller (Since the sine of the critical angle is equal to the reciprocal of the index of refraction, a greater index of refraction results in a smaller critical angle.) (frame 15)

14. 24°. Solution: $\sin \theta_{crit} = \dfrac{1}{2.24} = 0.413$ (frame 15)

21 Lenses and Instruments

Prerequisite: Chapter 20

We have seen how the phenomenon of specular reflection is effectively used in mirrors. The phenomenon of refraction is widely used in lenses. In this chapter we will explore the varied applications of refraction in lenses.

OBJECTIVES

After completing this chapter, you will be able to

- differentiate between converging and diverging lenses;
- define the focal length and focal point of a lens;
- differentiate between real and virtual images;
- relate image position to focal length and object distance for converging and diverging lenses;
- determine the magnification produced by a given set of conditions;
- explain the function of component parts of a camera—diaphragm, film, and lens;
- compare the operation of a camera to the operation of the human eye;
- indicate the type of lens needed to correct the two common eye defects;
- use a diagram to explain how a projector operates;
- identify the object and image for each lens in a microscope;
- determine the total magnification of a two-lens microscope given relative distances of objects and images;
- indicate the primary difference between a microscope and a telescope;
- specify the two major functions of a telescope and what parts of the telescope contribute to each;

- calculate the magnification of a telescope given the focal lengths of the two lenses;

- use the lensmaker's formula to calculate the focal length of a lens given the radii of curvature of the sides and the index of refraction (optional);

- calculate the position of an image, given focal length and object position for both converging and diverging lenses (optional).

1 LENSES

We saw in the last chapter (frame 19) that light changes direction (and separates into colors) when it goes through a prism. This bending occurs because the two sides of the glass are not parallel. The figure in this frame illustrates a glass lens with one ray of light going through it. If you draw normals to the surfaces of the glass where the ray crosses them, you can verify that the law of refraction is obeyed in each case.

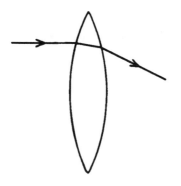

(a) In which of the cases here, X or Y, will the ray of light be bent *upward* by the prism? _____

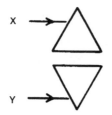

(b) What will happen to the ray of light about to strike the lens in figure Z?

_____

Answers:
(a) Y (The ray will bend as shown at right.)

(b) It will bend upward.

2 The figure in this frame shows a great number of rays passing through a lens. After they pass through, they come together, or converge. Lenses such as this—thicker at the center than at the edges—are called **converging lenses**. They are three-dimensional, appearing somewhat like that shape obtained by inverting one dinner plate on top of another. In fact, if the lens is made correctly (as is the one shown), the rays converge so that they cross at a single point. In this figure the

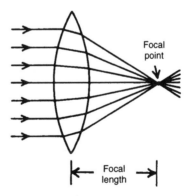

rays are parallel to one another and to the axis of the lens as they enter at the left. The point at which parallel rays converge after passing through a lens is called the **focal point** of the lens. The distance from the lens to the focal point is called the **focal length** of the lens.

The second figure in this frame shows the same lens as before, but in this case the rays are coming from the right.

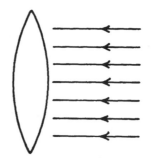

(a) Will the rays converge in this case or will they diverge (spread out)?

_____

(b) Where will the focal point be located? _____

Answers: (a) converge (Those at the top will be bent downward and those at the bottom will bend upward.);   (b) to the left of the lens

All lenses have two focal points, and the points are located at equal distances on each side of the lens.

3   Here we see two more converging lenses. Notice that the two lenses, although shaped differently, are both thicker at the center than at the edges. It is this feature that causes them to be converging lenses.

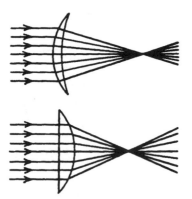

(a) If the light had struck the upper lens from the right instead of from the left, would it converge after passing through the lens? _____

(b) Which of the two lenses has the longer focal length? _____

Answers: (a) yes;   (b) the upper one

4 THE LENSMAKER'S FORMULA (OPTIONAL)
(Prerequisite: Chapter 20, frames 10, 11)

Two factors influence the focal length of a lens:

the curvature of the lens surfaces;

the index of refraction of the glass (or plastic).

The curvature is what makes a lens act differently from a pane of glass. The index of refraction determines the amount of bending when a ray of light passes from air into the lens (and then back out). The equation by which one can use the two factors to calculate the focal length of a lens is called the **lensmaker's formula**:

$$\frac{1}{f} = (n-1) \cdot \left(\frac{1}{r_1} + \frac{1}{r_2} \right)$$

where f is the focal length of the lens, n is the index of refraction, and r_1 and r_2 are the radii of curvature of the two lens surfaces.

The figure shows a lens with radii of curvature of 20 cm and 25 cm and with glass of index of refraction 1.52. Use the formula to calculate its focal length.

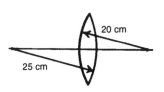

Answer: $f = 21.4$ cm.

Solution: $\dfrac{1}{f} = (1.52 - 1) \cdot \left(\dfrac{1}{20} + \dfrac{1}{25} \right)$

$= (0.52)(0.05 + 0.04)$

$= 0.0468$

$f = \dfrac{1}{0.0468}$

5 In the lens you just used, all of the values were positive. If one of the surfaces of the glass were curved in the other direction, its radius would be negative. For example, the right surface of the lens shown next is backward from the "standard" above. Consider its radius to be *negative* 30 cm. What is the focal length of this lens? _____

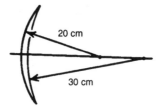

Answer: $f = 114$ cm

Solution: $\dfrac{1}{f} = (1.52 - 1) \cdot \left(\dfrac{1}{20} + \dfrac{1}{-30} \right)$

$= (0.52)(0.05 - 0.033)$

$= 0.0088$

6 IMAGES

In this figure, two sets of parallel rays are seen coming from the left toward a lens. They could represent rays from the top and the bottom of a very distant object. (If an object is distant enough, the light rays from any point on it are essentially parallel.) Note that the rays from each point on the object again converge to a point after passing through the lens, and that this point is at a distance equal to the focal length of the lens. One set of rays focuses below the focal point, however, because its incoming rays were not parallel to the axis of the lens.

If the top of the distant object is a red lightbulb, a red spot will appear on a piece of paper held where the rays converge. If the bottom of the object is a green lightbulb, a green spot will appear on the paper above the red spot. And the rays from points on the object between the top and bottom will appear on the paper also. In fact, an image of the object will be seen on the paper.

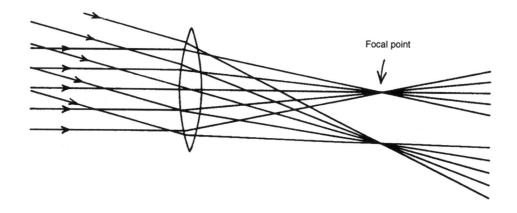

Focal point

(a) Will the positions of the red and green spots on the paper be inverted from the positions of the lightbulbs? _____

(b) Is the distance between the red and green spots on the paper greater or less than the distance between the red and green lightbulbs? _____

Answers: (a) yes;   (b) less

7   This figure shows a lens with an object close to it rather than at a great distance. In this case, the image will not appear at the focal plane (the vertical plane through the focal point), but will be more distant from the lens. With a nearby object, the rays are spreading out (diverging) when they reach the lens and the lens is unable to converge them as close to it as it can parallel rays. As in the case of the very distant object, the image is upside down, or inverted.

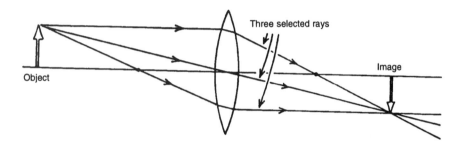

(a) If an object is moved farther from a lens, what happens to its image?

_____

(b) If the object is extremely far away (at "infinity"), where is the image located?

_____

Answers: (a) It gets closer to the lens.   (b) on the focal plane

8   We call the images we have been considering **real images**. The light actually passes through the position of the image, and if a piece of paper is put at that position, the image will appear on the paper. We saw in frame 3 of the last chapter the other type of image, the **virtual image**. Look back at the figure there and recall that the image formed by a flat mirror is located behind the mirror. Therefore, no light is actually at the position of the image. The light that bounces from the mirror only *appears* to come from the image. Such images are called virtual images.

(a) If a piece of paper were put at the position of the image of a plane mirror, would the image appear on the paper? _____

(b) When you watch a movie on a screen, are you seeing a real or a virtual image?

_____

(c) When you look at yourself in a mirror, is the image you see real or virtual?

_____

Answers: (a) no;   (b) real (Because light actually comes from the position of the image on the screen.)   (c) virtual

9 │ MAGNIFICATION

Magnification is defined as the ratio of the height of the image to the height of the object. Thus, if the magnification is 2 (sometimes written as 2×), the image is twice as tall as the object. In equation form this is

$$M = \frac{I}{O}$$

where M is the magnification, I is the image size, and O is the object size.

In the figure in frame 7 the image was slightly smaller than the object, so the magnification was slightly less than one. Now look at the figure in this frame. The image is 4 times taller than the object. The magnification is 4.

There is something else important to be seen in the figure. The image is farther from the lens than is the object. In fact, it is 4 times farther from the lens. In general, the relationship between the object and image size, and the object and image distance, is as follows:

$$\frac{I}{O} = \frac{q}{p}$$

where q is the image distance from the lens and p is the object distance from the lens.

Thus, magnification can be calculated either from the ratio of heights or from the ratio of distances to the lens.

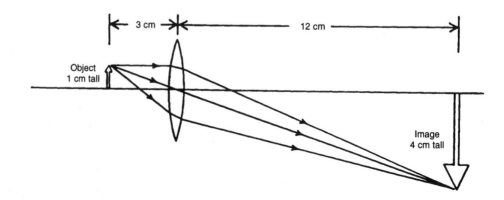

(a) Calculate the magnification using the distances of image and object from the lens for the situation shown in the figure in this frame. _____

(b) If the object were placed twice as far from this lens, would the magnification be the same? (Hint: The image would not remain in the same position.)

_____

Answers: (a) 4; (b) no (Recall that as the object moves farther away, the real image moves closer.)

---

10 IMAGE CALCULATIONS (OPTIONAL)

The relationship between the focal length f of a lens, the object distance p, and the image distance q is:

$$\frac{1}{f} = \frac{1}{p} + \frac{1}{q}$$

All quantities are positive when a converging lens is used and the object and image are on opposite sides of the lens (which is the only case we have considered so far).

Suppose you have a lens with a focal length of 12 cm and you locate an object 20 cm from it. Where will the image be? _____

Answer: 30 cm from the lens (on the opposite side)

Solution:
$$\frac{1}{12} = \frac{1}{20} + \frac{1}{q}$$
$$0.0833 = 0.05 + \frac{1}{q}$$
$$0.0333 = \frac{1}{q}$$

---

11 VIRTUAL IMAGES: THE MAGNIFIER

When an object is very distant from the lens, the image is located at the focal point, or on the focal plane. As the object moves closer, the image moves farther from the lens. Finally, when the object is put at the focal point, the image is located very far away (at "infinity," actually). What happens when the object is placed inside the focal point? In this case the lens is unable to bring the rays back to converge to a single point. Refer to the figure in this frame.

In this case the rays from the top of the object, after going through the lens, seem to be coming from a point farther back on the left. In fact, an image of the object is formed there. The rays of light do not actually go through the image, so

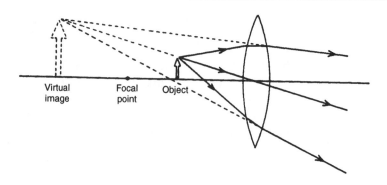

the image is a virtual one. If a piece of paper were put at the location of this virtual image, an image would not appear on the paper. We can say that an image is there, however, because if the rays going toward the right enter an eye (or a camera) the object seems to be located where the image shows in the diagram.

(a) Which is larger in the case we are discussing, the object or the image?

_____

(b) Which is farther from the lens? _____

(c) In what way is this image similar to the image in a plane mirror? _____

_____

Answers: (a) the image; (b) the image; (c) Both are virtual images.

12 Refer to the figure in this frame. Notice that the virtual image is larger than the object again. (This will always be the case when the object is inside the focal point of a converging lens.) Imagine that you are looking at the situation from the right (so that the rays pass into your eyes). A lens used in this manner is often called a magnifying glass, or a magnifier. The relationship between sizes and distances holds for this case just as for the real images studied before.

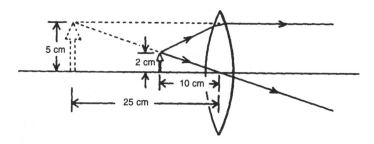

(a) Calculate the magnification in the figure using the ratio of image size to object size. _____

(b) Calculate the magnification using the ratio of distances—image distance to object distance. _____

Answers: (a) 2.5;   (b) 2.5

---

13 CALCULATIONS OF VIRTUAL IMAGE DISTANCES (OPTIONAL) (Prerequisite: frame 10)

The equation relating focal length, object distance, and image distance, $1/f = 1/p + 1/q$, also holds in the case of virtual images. The image distance is negative, however, since it is on the opposite side of the lens from what it is in the "standard" situation.

(a) Calculate the focal length of the lens in the figure in frame 12 using the object and image distances given there. _____

(b) Suppose an object is located at a distance of 6 cm from a lens with a focal length of 10 cm. Where will the image be located? _____

Answers:

(a) 16.7 cm. Solution: $\dfrac{1}{f} = \dfrac{1}{10} + \dfrac{1}{-25}$ (Note the minus sign.)

$$= 0.10 - 0.04$$
$$= 0.06$$

(b) –15 cm (The minus sign indicates that the image is on the same side of the lens as the object.)

Solution: $\dfrac{1}{10} = \dfrac{1}{6} + \dfrac{1}{q}$

$$0.1 = 0.167 + \dfrac{1}{q}$$

$$\dfrac{1}{q} = -0.067$$

---

14 DIVERGING LENSES

The first figure in this frame shows parallel rays of light striking a lens that is thinner at the center than at the edge. Such a lens diverges light and is thus called a **diverging lens**. The light that emerges from the lens acts as if it comes from a point closer to the lens than it actually does. The point from which incoming parallel light seems to come after passing through a diverging lens is the focal point of the lens. As with converging lenses, the distance from the focal point to the lens is the focal length.

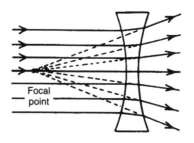

The next figure shows the formation of an image by a diverging lens. The image formed by such a lens is always smaller than the object.

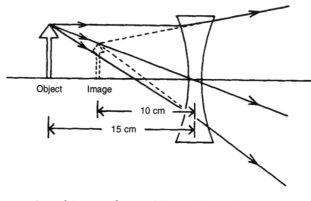

(a) Is a real or a virtual image formed by a diverging lens? _____

(b) What type lens can be used as a magnifier? _____

Answers: (a) virtual; (b) converging

15 DIVERGING CALCULATIONS (OPTIONAL) (Prerequisite: frame 10)

Image and object distances for diverging lenses are calculated as for converging lenses. But now we have another sign change, since the focal length of a diverging lens is negative. Use the given object and image distances to calculate the focal length of the lens of the second figure in frame 14. _____

Answer: −30 cm

Solution: $\dfrac{1}{f} = \dfrac{1}{15} + \dfrac{1}{-10}$ (image on left, thus minus sign)

$= -0.0333$

16 | THE CAMERA

The basic camera is perhaps the simplest possible use of a lens. The figure here is a diagram of a box camera. An image of the distant object is formed on the film.

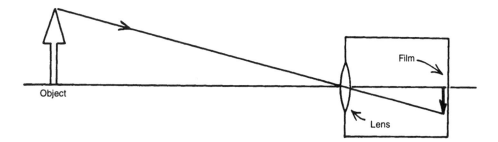

You will recall that when the object distance is changed, the image distance is likewise changed. The simplest cameras do nothing about this problem. Their lens is set so that a fairly distant object is in good focus. A better camera has an adjustable lens-to-film distance so that it can be focused for objects at various distances. In addition, a good camera contains an adjustable opening—called a diaphragm—that can be set to allow different amounts of light to reach the film. In effect, the diaphragm adjusts to make the lens larger or smaller.

(a) To take a photo of a distant object, what should be the lens-to-film distance?

_____

(b) Is the image on the film right side up or upside down? _____

(c) When an adjustable camera is used to take a close-up, is the lens moved closer to or farther from the film? _____

(d) How is the diaphragm changed to take a photo in dim light? _____

_____

Answers: (a) the focal length of the lens; (b) upside down; (c) farther from (Recall from frame 7 that when the object is moved closer, the image moves farther from the lens.); (d) It is opened, to allow more light in.

17 | THE EYE

The next figure illustrates the human eye. It operates very similarly to the camera (or vice versa, since the eye came first). Light passes through the cornea and lens to focus an image on the retina. The iris controls the amount of light entering the eye.

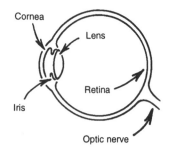

What part of the camera corresponds to each part of the eye listed below?

(a) iris _____

(b) retina _____

(c) lens and cornea _____

Answers: (a) diaphragm;   (b) film;   (c) lens

18   The adjustable camera allows you to focus on objects at different distances by moving the lens. The eye, however, has a lens that can be made to change shape. When the muscles around it contract, the lens assumes a more curved shape and is therefore more converging. Such shortening of the focal length of the lens is necessary in order to bring a nearby object into focus. This is because it must reconverge the greatly diverging rays from the nearby object.

(a) Would the eye lens be more or less curved for viewing distant objects? _____

(b) What type image, real or virtual, is formed on the retina of the eye? _____

Answers: (a) less curved;   (b) real

19   The most common vision defect for young people is nearsightedness (myopia). This is caused either by the eyeball being too long or the lens being too powerful (too converging). The result is that the image is brought to a focus in front of the retina. (See figure A on the next page.) Such a defect is corrected by placing a diverging lens in front of the eye in order to spread the light and make it focus farther from the lens. This is shown in figure B.

Farsightedness (hyperopia) is the condition in which the image is (or would be) formed behind the retina. This is caused either by an eye that is too short or by a lens that is not converging enough. This condition is most common in middle and later ages, when the lens becomes somewhat stiff and the muscles are unable to squeeze it down into greater curvature. Figures C and D illustrate this defect and its correction.

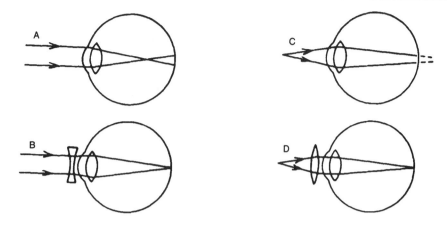

(a) Suppose the lens of figure B were used for the eye condition in C. Where (approximately) would an image be formed? _____

(b) What type lens is used to correct nearsightedness? _____

(c) If the fault of the eye is that its lens is too thin, what type eyeglass lens corrects this? _____

Answers: (a) farther behind the retina; (b) diverging; (c) converging (Since it is thicker at the center than at the edges.)

20 THE PROJECTOR

The projector is, in some ways, the reverse of the camera. The figure here shows a frame of movie film (or a slide) being illuminated by a light and the image being projected onto a distant screen. The screen is at a great distance compared to the focal length of the lens, while the film is placed very close to the focal point. A lens may be placed between the projector lamp and the film to concentrate light on the film, or a curved mirror may be placed behind the lamp to accomplish the same effect. In addition, projectors include a shield to block the light while changing slides on a slide projector, or while the movie film is going from one frame on the

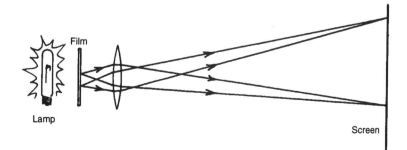

film to another. This shield would be located just to the left of the film in the figure.

(a) What type lens is used to project the image onto the screen? _____

(b) If the image on the screen were out of focus, what components in the figure could be adjusted to bring it into focus? _____

_____

Answers: (a) converging; (b) The distance between film and lens would be changed. (In fact, the distance between the lens and screen could be changed instead, but this is usually impractical.)

---

21 THE MICROSCOPE

Until the invention of the microscope, the simple magnifier was the only instrument available to see very tiny objects. The figure in this frame shows the optical system of a microscope. The lower lens, the **objective**, has a very short focal length, and the object to be viewed is placed just beyond its focal point. This means that the image formed by the lens will be relatively far from the lens, and that it will be much larger than the object (remember from frame 9 the relationship between sizes and distances from the lens).

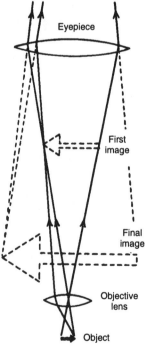

The upper lens, the **eyepiece**, uses the image formed by the objective as its object. The eyepiece, then, is acting as a simple magnifier. From the figure you can see that the final image is not only much enlarged, but is inverted in orientation from the original object.

The magnification produced by a microscope is the product of the magnification of each of the lenses.

(a) If the image formed by the objective is 25 times as far from the lens as is the object, what magnification does this lens produce? _____

(b) If the final image is 10 times as far from the eyepiece as is its object (the image formed by the objective), then what is the magnification of the microscope as a whole? _____

Answers: (a) 25 (see frame 9);   (b) 250 (10·25)

22 THE TELESCOPE

The major difference between the function of a telescope and a microscope is that the object being viewed with a telescope is at a great distance while for a microscope it is close. The figure here shows the optical system of a refracting telescope.

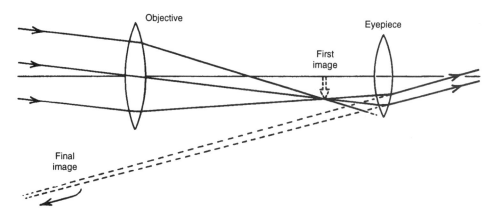

Since the object is very distant, the incoming rays are parallel. The first lens (the objective again) produces an image on its focal plane. Then, just as with the microscope, the eyepiece is used as a magnifier to view the image.

In normal use of a telescope, the eyepiece is located at such a distance from the image formed by the objective that the final image is very far away—at infinity. It is thus meaningless to compare object and image distances to calculate magnification. What is important when one looks through a telescope at the moon or a planet is the relative angle taken up by the rays as they enter the eye. It turns out that magnification, when defined in this way, is equal to the focal length of the objective lens divided by the focal length of the eyepiece. A telescope has a single large objective lens, so in order to change the telescope's magnifying power, the viewer changes eyepieces. To get a high magnification, does one use an eyepiece with a long or a short focal length? _____

Answer: short $\left(\text{because magnification} = \dfrac{f_{\text{objective}}}{f_{\text{eyepiece}}} \right)$

---

23   Since eyepieces of extremely short focal length can be made, there is theoretically no limit to how powerful a telescope can be made. Practically, however, telescopes do have a limit to power. Even on the clearest night, the atmosphere between us and the stars blurs the image if too much magnification is used. Magnification beyond a certain point is worthless—it just magnifies the fuzziness of the image. A telescope is also limited by the wavelength of the light, which will be discussed in the next chapter.

Telescopes are used not just for magnification, but more importantly to gather as much light as possible from dim stellar objects so that we can see them. For this reason, telescopes are constructed with large objective lenses. The largest objective lens in use on a refracting telescope is 40 inches in diameter (at the Yerkes Observatory).

(a) What is the advantage of a telescope with a large objective lens? _____

_____

(b) Why is greater magnification not always desirable? _____

_____

Answers: (a) It gathers more light.   (b) There is a limit to the clarity of the image.

SELF-TEST

1. How do converging lenses and diverging lenses differ in their appearance and in what they do to a beam of parallel light? _____

_____

_____

2. How does the focal point for a converging lens differ from the focal point for a diverging lens? _____

_____

_____

3. Suppose a converging lens forms an image of a very distant object. Where is the image located? _____

4. As a distant object moves closer to a lens, what happens to the position of the image? _____

_____

5. What is the relationship between magnification and image and object distances? _____

_____

6. An object is 23 cm from a lens, and the magnification produced by the lens is 3. Where is the image located? _____

7. What is the difference between real and virtual images? _____

_____

_____

8. Compare the sizes of the object and image when a diverging lens is being used. _____

9. How does the eye accommodate (focus on objects at different distances)?

   _____

10. What are the two most common vision defects? What type of lens is used to correct each? _____

11. What forms the object that is viewed by the eyepiece of a microscope? _____

   _____

12. Why is a large objective lens desirable for a telescope? _____

   _____

13. What is the magnification of a telescope with an objective of focal length 1.0 meter (100 cm) when an eyepiece of focal length 2.0 cm is being used?

   _____

14. If the objective of a microscope magnifies the object 20 times and the eyepiece produces a magnification of 15, what is the magnification of the microscope?

   _____

15. *Optional.* The lens diagrammed below is made of glass with an index of refraction of 1.6. What is its focal length? _____

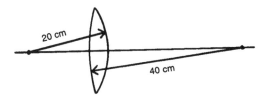

16. *Optional.* If an object is placed 25 cm in front of the lens of the last question, where will the image be located? _____

17. *Optional.* What is the focal length of the lens shown below if the index of refraction of the glass is 1.52? _____

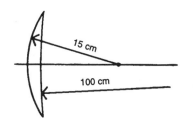

18. *Optional.* If an object is located 42 cm to the left of the lens of the last question, where will the image be located? _____

19. *Optional.* Suppose one wishes to use a lens of focal length 20 cm as a simple magnifier and have the image located 25 cm from the lens. Where must the object be placed? _____

20. *Optional.* A lens has a focal length of –20 cm. What type of lens is it?

_____

ANSWERS

1. A converging lens is thicker at the center than at the edges, while a diverging lens is the opposite. A converging lens brings parallel rays of light to a single point. A diverging lens spreads the rays so that they appear to have come from a single point behind the lens. (frames 2, 14)

2. The focal point of a converging lens is that point to which incoming rays that are parallel to one another and to the axis of the lens will converge after passing through the lens. For a diverging lens, the focal point is the point from which such rays seem to be coming. (frames 2, 14)

3. at the focal point (or, to be more precise, on the focal plane) (frame 6)

4. As the object approaches, the image recedes from the lens. When the object reaches the focal point, the image is at infinity. (frame 7)

5. Magnification is equal to image distance divided by object distance. (frame 9)

6. 69 cm from the lens (magnification = q/p, so $q = p \cdot$ magnification = 23·3) (frame 9)

7. A real image is one that is actually formed by light passing through the image position. In the case of a virtual image, the light appears to come from the image, but it does not actually reach the image. (frame 8)

8. The image is always smaller than the object. (frame 14)

9. The lens of the eye changes shape. (frame 18)

10. myopia (nearsightedness, corrected by a diverging lens) and hyperopia (farsightedness, corrected by a converging lens) (frame 19)

11. The object for the eyepiece is formed by the objective; it is the image formed by the objective. (frame 21)

12. It gathers as much light as possible so that dim objects may be seen. (frame 23)

13. 50 (magnification = $f_{objective} / f_{eyepiece}$ = 100 cm/2 cm) (frame 22)

14. 300 (the product of the two magnifications) (frame 21)

15. 22.22 cm. Solution:
$$\frac{1}{f} = (n-1) \cdot \left(\frac{1}{r_1} + \frac{1}{r_2} \right)$$
$$= (1.6-1) \cdot \left(\frac{1}{20} + \frac{1}{40} \right)$$
$$= 0.045 \quad \text{(frame 4)}$$

16. 200 cm from the lens. Solution:

$$\frac{1}{f} = \frac{1}{p} + \frac{1}{q}$$

$$\frac{1}{22.22} = \frac{1}{25} + \frac{1}{q}$$

$$\frac{1}{q} = 0.0050 \quad \text{(frame 10)}$$

17. 33.9 cm. Solution:

$$\frac{1}{f} = (n-1) \cdot \left(\frac{1}{r_1} + \frac{1}{r_2}\right)$$

$$= (1.52 - 1) \cdot \left(\frac{1}{15} + \frac{1}{-100}\right)$$

$$= 0.0295 \quad \text{(frame 5)}$$

18. 176 cm from the lens. Solution:

$$\frac{1}{f} = \frac{1}{p} + \frac{1}{q}$$

$$\frac{1}{33.9} = \frac{1}{42} + \frac{1}{q}$$

$$\frac{1}{q} = 0.00569 \quad \text{(frame 10)}$$

19. 11.1 cm from the lens. Solution:

$$\frac{1}{f} = \frac{1}{p} + \frac{1}{q}$$

$$\frac{1}{20} = \frac{1}{p} + \frac{1}{-25}$$

(The minus sign appears because the lens is being used as a magnifier, so the image is on the same side of the lens as the object.)

$$\frac{1}{p} = 0.090 \quad \text{(frame 13)}$$

20. diverging (Because the focal length is negative.) (frame 15)

22 Light as a Wave

Prerequisite: none, but Chapters 10 and 12 are suggested
In the beginning of our study of light (Chapter 18) we considered the nature of light—is it wave or particle? Chapter 19 focused on the particle nature of light, but the phenomenon of refraction (studied in Chapters 20 and 21) was explained using a wave model. In this chapter we will study additional phenomena that rely on a wave theory of light.

OBJECTIVES

After completing this chapter, you will be able to

- define diffraction;

- describe how light bends around corners;

- indicate the effect of constructive and destructive interference of light waves;

- explain the cause of the pattern that results from double slit interference;

- state the relationship between interference by a diffraction grating and double slit interference;

- explain the colors of soap bubbles in terms of wave theory;

- state the conditions necessary to observe diffraction of light;

- indicate how crossed polarizers reduce the amount of light that passes through them;

- use the double slit interference equation to determine the location of bright spots that result from double slit interference (optional);

- state the effect of wavelength on the double slit interference pattern (optional);

- use the diffraction grating equation to determine the location of bright spots in the diffraction pattern (optional).

1 DIFFRACTION

In our everyday living we assume that light travels in a straight line. We are aware of exceptions—when light goes through a lens or prism—but most of us assume that when light travels through still air, it goes straight. We align objects by sighting along them. We aim a gun by sighting along it. These activities depend upon light not bending its path. But light does bend around corners! It only bends a little, however, and we will notice it only if we know where to look.

The best way to see the bending of light around corners (called "**diffraction**") is to use a razor blade to cut a short slit in a piece of paper, and then look through the slit at a distant source of light. Hold the slit close to your eye and open and close the slit by flexing the paper. You will see the light apparently spread out perpendicular to the slit. The figure in this frame illustrates light passing through a slit that is perpendicular to the plane of this page. The light spreads out before entering the eye.

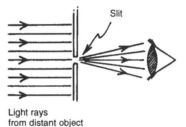

Light rays
from distant object

(a) Is diffraction the result of light going from one material into another like refraction? _____

(b) Is diffraction the result of the bouncing of light from a surface (like reflection)? _____

Answers: (a) no;   (b) no

2

The phenomenon of diffraction is most easily explained by a wave model of light. If light consisted of particles, we would expect it to pass on through an opening without spreading. Waves, however, do show spreading after passing through an opening. For example, sound waves spread out after they pass through a door or window, and water waves will do the same thing after they pass through small openings. If you try the latter by using two sticks to make a $\frac{1}{4}$- to $\frac{1}{2}$-inch opening for water waves in a pan, you will find that the smaller the opening the more the spreading. This corresponds to what we see in the case of light—the smaller the opening, the greater the diffraction effect.

In addition, sound waves and water waves bend around objects in their path. If light did this on a large scale, there would be no shadows, but if a very small object is placed in a beam of light, such bending can be observed. (It is best observed with a laser—the bright light of the laser being easy to see after it curves around the object.)

(a) What term refers to the bending of light around corners? _____

(b) What amount of diffraction would result if light waves were particles instead?

_____

(c) Is the bending more pronounced when the opening is large or small?

_____

Answers: (a) diffraction;   (b) none (because we expect particles to continue in a straight line);   (c) small

3 DOUBLE SLIT INTERFERENCE*

The figure in this frame shows waves that spread from a source (L) and strike two slits (M and N). After passing through each slit, the light again spreads out, so that light from one of the slits overlaps light from the other. The line from B to F represents a screen upon which the light falls. In the region between the slits and the screen, we represent the crests of waves as solid arcs and the troughs as dashed arcs. Notice that along a line that starts between M and N and goes to point D, the crests of waves that left point M overlap with the crests of waves that left point N. And along this same line, troughs from M overlap troughs from N. The locations of these overlaps are marked with small circles. The overlapping of crest-on-crest and trough-on-trough tells us that a strong wave hits point D, and a bright spot will appear at D. The same thing happens along lines from MN to point B and to point F, so that points B and F on the screen will also appear bright.

Now look at a line from MN to point C on the screen. Along this line, crests from point M overlap with troughs from point N, and troughs from M overlap crests from N. The points of overlap are marked here with open circles. The overlapping of crests and troughs means that the waves cancel along these lines. This causes a cancellation of light—dark spots—at points C and E.

At points where the two waves add so as to reinforce each other we say that **constructive interference** is taking place. This is occurring at points B, D, and F. At points such as C and E, where the waves cancel one another, we say that there is **destructive interference.**

Referring to the figure, indicate whether there is constructive or destructive interference for each of the points.

_____

*You may want to look over frames 3–5 in Chapter 12 before studying this. Those frames deal with a similar case for sound waves.

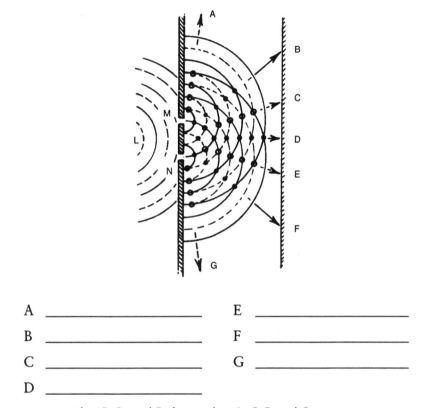

A _____   E _____

B _____   F _____

C _____   G _____

D _____

Answers: constructive: B, D, and F; destructive: A, C, E, and G

4

To observe double slit interference in practice, the slits must be very small, like those you used to observe diffraction. In addition, the slits must be very close together, about 0.1 mm or less. Since a regular light bulb does not produce intense enough light to pass through such narrow slits and still illuminate a screen visibly, one normally observes "double slit interference" either by substituting the retina of the eye for the screen (by holding the slits just in front of the eye) or by using a laser as a source of light.

Suppose you manage to get a double slit interference pattern on a screen. What pattern will you see? _____

Answer: bright and dark spots across the screen

5

Optional. Several variables determine how much distance there will be between consecutive bright spots on a screen illuminated by double slit interference. Perhaps the most obvious variable is the distance from the slits to the screen (distance D in the figure of this frame). Since the pattern spreads from the slits, we would expect that the larger that distance D is, the farther apart will be the bright

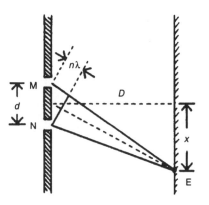

spots on the screen. The other variables involved are the wavelength of the light (λ) and the separation of the slits (d).

The figure shows lines from each of the slits to a spot (E) on the screen. If the spot is to be bright, the distance from M to E must be greater than the distance from N to E by exactly one wavelength (or two wavelengths, or three, and so on). This extra distance is labeled in the figure as $n\lambda$ (n indicating 1, 2, 3, and so on). The small triangle with $n\lambda$ as one of its sides is similar to the large (dashed-line) triangle with x as one of its sides. Thus, making a slight approximation, we can write a ratio of length of sides:

$$\frac{n\lambda}{d} = \frac{x}{D}$$

Suppose we have two slits 0.1 mm apart and shine light through them on to a screen 5 meters distant. The first bright spot beyond the center one is measured to be 2.0 cm from the center spot.

(a) What is the wavelength of the light used? (Hint: First put all variables in meters.) _____

(b) The second bright spot will be how far from the center spot? (This is found by using $n = 2$.) _____

Answers:
(a) $4 \cdot 10^{-7}$ meters (or 400 nanometers)

Solution: $\dfrac{n\lambda}{d} = \dfrac{x}{D}$

$\lambda = \dfrac{xd}{Dn}$

$= \dfrac{(0.02)(0.0001)}{(5)(1)}$ (all distances in meters)

(b) 0.04 meters (or 4 cm)

Solution: From the equation: $x = \dfrac{\lambda D n}{d}$

$$= \dfrac{(4 \cdot 10^{-7})(5)(2)}{0.0001}$$

6 THE DIFFRACTION GRATING

The double slit interference pattern is different for different wavelengths, and it could be used to analyze light (similar to the way a prism is used), but the pattern is normally fairly dim and the bright spots are very close together. Making the slits closer together spreads out the pattern, putting the bright spots farther apart. But as the slits are made closer together, they must also be made smaller. Since less light then goes through them, the pattern will be dimmer. This situation can be rectified by including more than two slits.

In a **diffraction grating** there are thousands of slits per centimeter, so the slits are less than a thousandth of a centimeter apart. (They may be less than a ten-thousandth of a centimeter apart.) Such a grating is made by mechanically scratching a piece of glass with extremely fine lines. The figure in this frame shows how the slits would appear under high magnification. Where there are scratches, the glass is rough and the light cannot pass through. Thus, the light passes between the scratches, so that these locations are equivalent to the slits in the previous two figures. Diffraction gratings spread the light so much that the first bright spot may be at an angle of 30 degrees or more from center. Diffraction gratings are used in place of prisms to analyze light, since each color has its bright spot at a different angle.

The diffraction grating discussed above works by transmission of light. Diffraction can also be observed in reflected light, however. If you look at an angle across a CD (a compact disk), you can see color in the reflection. This color

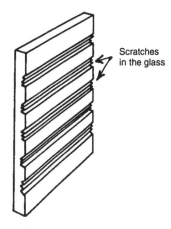

Scratches
in the glass

results from interference of the light that bounces from the various dots on the CD, and is the result of essentially the same process as occurs with the standard diffraction grating.

(a) What is it about a diffraction grating that causes its pattern to be spread out more than the double slit pattern? _____

(b) How is it that the diffraction grating pattern is brighter than the double slit pattern? _____

Answers: (a) The slits are closer together.   (b) The diffraction grating has many slits for the light to come through.

7

Optional. The diffraction grating spreads the light at a far greater angle than simple double slits. For this reason we write the equation in terms of the angle of spread as follows:

$$n\lambda = d \cdot \sin \theta *$$

where d is again the distance between adjacent slits, and the angle (θ) is shown in the figure here.

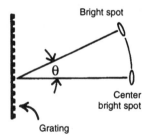

Use a calculator to determine the angle of the second bright spot produced by light of wavelength 400 nanometers (1 nanometer = 10^{-9} meters) passing through a grating with 8000 lines per centimeter. (Hint: You must first find the distance d between slits.) _____

Answer: 40°. Solution: If there are 8000 lines per centimeter, the distance between adjacent lines is 1/8000 cm, or $1.25 \cdot 10^{-4}$ cm, or $1.25 \cdot 10^{-6}$ m. Putting all distances in meters:

$$\sin \theta = \frac{n\lambda}{d}$$
$$= \frac{(2)(400 \cdot 10^{-9})}{1.25 \cdot 10^{-6}}$$
$$= 0.64$$

---

* "Sin" is the abbreviation for sine. If you have studied trigonometry, you can understand from the figure in frame 5 why the sine function appears here.

8 THIN FILM INTERFERENCE

In order to produce interference of light, a beam of light must somehow be broken into two beams and then recombined so that one of the separated beams travels farther than the other. We saw that this could be done by the use of double slits or a diffraction grating. Another example of interference is the production of the colors in an oil slick on a wet road. The figure in this frame illustrates a greatly magnified section of a portion of such an oil slick, which is a thin film of oil over the water that covers the road.

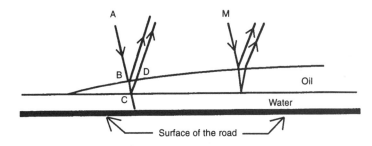

A light ray is shown coming from point A and striking the oil at B. When it hits the surface of the oil, some of the light is reflected from the surface, but some continues through it. When this latter portion of the light hits the surface of the water (at C), some of it is again reflected and some continues on to the road. We will concern ourselves here with the reflected portion. It goes back toward the upper surface of the oil where some of it emerges (at D) and recombines with the portion of the original beam that reflected from the surface upon first encountering it. These two waves overlap. (In the drawing, they do not seem to overlap. This is because we have exaggerated the thickness of the oil film so much. In addition, we show the light hitting the oil at a fairly large angle, while in practice this effect is seen for light striking the surface nearly perpendicular.)

Which part of the light has gone the greater distance? _____

Answer: The part that passed to the inner surface and bounced back.

9

Now suppose that the light that strikes the oil film is of a single wavelength. In this case, the extra distance that the one part of the beam travels might be just the right distance so that it is in phase with the original beam when they combine again. This will happen if the extra path distance is one wavelength, two wavelengths, or three wavelengths, etc.

(a) What condition will create constructive interference? _____

(b) What will be observed where constructive interference takes place?

_____

Answers: (a) It will happen when the greater distance is a multiple of the wavelength. (b) The spot will appear bright.

10   Continue to imagine that the light is of a single wavelength. Consider the beam that originates from point M in the figure. It strikes the oil film where the film is thicker. If this thickness is such that the extra path distance of the portions of the light that enters the oil is $\frac{1}{2}$ (or $1\frac{1}{2}$ or $2\frac{1}{2}$ or $3\frac{1}{2}$, and so on) wavelength, the two waves will interfere destructively upon recombining. That particular wavelength of light will therefore not be seen to reflect at that point on the oil.

(a) What condition will create destructive interference? _____

_____

(b) What will be observed where destructive interference takes place? _____

Answers: (a) It will happen when the greater distance is an odd multiple of a half-wavelength ($\frac{1}{2}$, $\frac{3}{2}$, $\frac{5}{2}$, etc.).   (b) The spot will appear dark.

11   In practice, oil films on highways are usually illuminated with white light, which contains all wavelengths of visible light.

(a) How does light of different wavelengths appear to the eye? _____

_____

(b) If white light (containing all wavelengths) strikes the oil film, would all colors be reflected at all points? _____

Answers: (a) as different colors;   (b) no (At some points, particular colors cancel out upon reflection and other colors interfere constructively.)

This explains why "rainbows" appear in oil slicks. If you push the oil around with your foot, you will change its thickness and change the colors that you see. A similar interference effect is observed in the thin wall of soap bubbles.

12   POLARIZATION

When water waves move across the surface of water, the primary motion of the water is up and down. By the very nature of the wave, they must vibrate vertically. If you vibrate the end of a rope or spring that is stretched between you and a wall, however, you can cause waves to vibrate vertically, horizontally, or at any angle in between. Light waves are similar. Figure X on the next page shows how

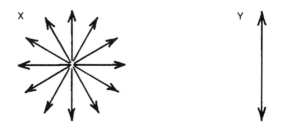

we might think of the vibration of a light wave as it is coming straight toward us. The arrows are used to represent vibration in all directions. Such a wave is termed **unpolarized.**

If unpolarized light is passed through a polarizer, all waves are absorbed except those that vibrate in one particular plane, such as illustrated in Figure Y.

(a) Now consider the next figure. Is the wave at A polarized or unpolarized?

_____

(b) What does the polarizer do? _____

_____

(c) The polarizer at D is at a 180° angle to the one at B. What will be its effect on the light polarized by B? _____

_____

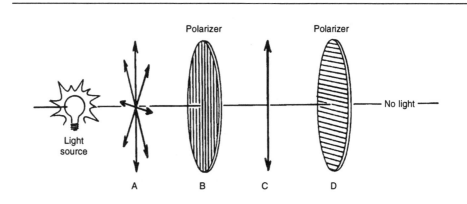

Answers: (a) unpolarized;   (b) It allows only waves vibrating in the vertical plane to get through. (c) It doesn't allow vibrations in the vertical plane through at all, so no light gets through.

13

The last figure showed the use of two polarizers to eliminate light. The figure in this frame shows another method. Here light reflects from a shiny surface. Such a surface polarizes light (at least partially) in the horizontal plane. In the figure, the reflected light is shown going toward a polarizer that allows vertical vibrations to pass. Thus, if the reflection results in perfect polarization, none of the reflected light gets through. This is the secret of polarizing sunglasses. The glare that results from reflection from smooth surfaces such as car bumpers and water surfaces is at least partly eliminated by such glasses. To see this effect, rotate polarized glasses in front of your eyes as you look through them at bright reflections from various objects.

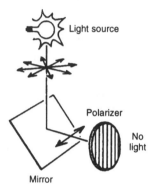

How is it that polarized sunglasses don't prevent you from seeing at all?

_____

Answer: Not all of the light is polarized upon reflection.

SELF-TEST

1. Why do we not see diffraction of light as it passes through doorways? _____

   _____

   _____

2. Why is light canceled at the dark spots in double slit interference? _____

   _____

3. Why does a diffraction grating result in the bright spots being so much farther apart than the bright spots produced by double slits? _____

   _____

4. Refer to the figure in frame 8. How much farther must the beam that enters the oil film travel if a bright area is to be seen? _____

   _____

5. Why are different colors seen at different places on an oil slick? _____

   _____

6. How does polarized light differ from unpolarized light? _____

_____

7. *Optional.* Suppose double slits are spaced $1.33 \cdot 10^{-4}$ meters apart and light of 500 nanometers wavelength shines through them. What distance from the center spot will the second bright spot appear on a screen 4 meters away?

_____

8. *Optional.* A grating with 12,000 lines per centimeter is used with light of wavelength 650 nanometers. What is the angle of the first bright spot? _____

9. *Optional.* What effect on the double slit pattern will result from an increase in the wavelength of light? _____

ANSWERS

1. Diffraction of light only occurs to a noticeable degree when the opening through which the light goes is very small. (frames 1, 2)

2. The troughs of one of the beams meet the crests of the other and vice versa. Crest cancels trough, and a dark spot results. (frame 3)

3. The slits of a diffraction grating are much closer together than it is practical to make double slits. (frame 6)

4. It must travel a whole number of wavelengths farther. (frame 8)

5. At any given place, only one wavelength will show perfect constructive interference. All other wavelengths will interfere destructively to some degree. (frames 8, 9)

6. In polarized light, the wave vibrates in only one plane, while unpolarized light has waves vibrating in all directions perpendicular to the direction of travel. (frame 12)

7. 0.03 meter (or 3 cm)

   Solution: $\dfrac{n\lambda}{d} = \dfrac{x}{D}$

   $x = \dfrac{n\lambda D}{d}$

   $= \dfrac{(2)(500 \cdot 10^{-9})(4)}{1.33 \cdot 10^{-4}}$ (in meters) (frame 5)

8. 51°. Solution: $\dfrac{1}{12,000} = 8.33 \cdot 10^{-5}$, which is the grating space in centimeters. This is $8.33 \cdot 10^{-7}$ meters.

   $$\sin \theta = \frac{n\lambda}{d}$$

   $$= \frac{(1)(650 \cdot 10^{-9})}{8.33 \cdot 10^{-7}}$$

   $$= 0.780$$ (frame 7)

9. The spots will get farther apart. This follows from the equation:

   $$\frac{x}{D} = \frac{n\lambda}{d}$$

   As the wavelength (λ) gets larger, so does distance x, the distance between slits. (frame 5)

23 Color

No prerequisites

Visible light is electromagnetic radiation with wavelengths between 400 nanometers and 700 nanometers. (A nanometer is 10^{-9} meters.*) Within this range, a difference in wavelength appears to us as a difference in color.

OBJECTIVES

After completing this chapter, you will be able to

- state the relationship of color to wavelength;
- state which colors have the longest and which have the shortest wavelength;
- name the additive primary colors;
- specify the result when given colors are combined additively;
- specify the method of color combination used on a television screen;
- name the subtractive primary colors;
- specify the result when given colors are combined subtractively;
- explain why making accurate predictions of the colors that result from combining paints in real life is not as simple as described in this chapter.

1 THE VISIBLE SPECTRUM

The figure in this frame shows the order of the colors in the visible spectrum, from violet light of shortest wavelength to red at the longest wavelength. It is important to realize that the range indicated for each color in the figure is arbitrary. Red fades gradually into orange, and different people might put the

---

*See Appendix I for a discussion of powers of ten.

Violet   Blue   Green   Yellow   Orange   Red

Short wavelength   Long wavelength

dividing line between the two colors in different places. In fact, the number of colors that you might list across the spectrum is arbitrary. You might list aqua as a separate color between blue and green, or indigo between blue and violet.

Which has the shorter wavelength, a blue-green or a yellow-green color?

_____

Answer: blue-green

2

White light contains all wavelengths of the spectrum, so it contains all colors. Yet, when white light shines on a banana, the light that enters your eye from the banana is yellow. This is because the surface of the banana absorbs some colors from the white light and reflects mostly yellow light.

What causes an unripe banana to appear green? _____

_____

Answer: The surface absorbs other colors and reflects green.

3

Although the color green is shown as a small portion of the visible spectrum, if light of the entire central third of the spectrum strikes the eye, we see the light as being green. Thus, for convenience, we may divide the spectrum into thirds and name them as shown in the figure here. The long-wavelength third appears red to us, although it actually includes wavelengths that by themselves appear orange. The short-wavelength third appears blue.

Now suppose that we have a wall that normally appears white (because it reflects all colors of light). If we shine light that contains only the central portion of the spectrum on this wall, the wall will appear green. What if we shine green

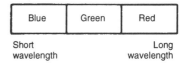

light on this wall at the same time that we are shining red light on it? In this case, two-thirds of the spectrum is being reflected into our eye from the wall. (See the next figure, part A.) This two-thirds of the spectrum appears yellow to us. (You will note that on the figure in frame 1, which shows all colors of the spectrum, yellow appears in the center of this range.) Refer to parts B and C of the figure.

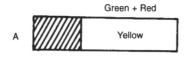

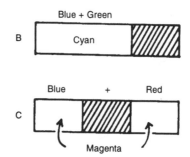

(a) If green and blue light are shined on a white wall simultaneously, what color do we see? _____

(b) If red and blue light are shined on the white wall, what color do we see?

_____

Answers: (a) cyan (This blue-green color is sometimes called aqua.); (b) magenta (This is a purplish red.)

4

In the previous frame, we formed new colors by shining certain colors on a white wall. In this way we were actually adding the colors together. This process is called an **additive process**. On the screen of a color television set are small dots (or rectangles) of the three additive primary colors, small enough that we cannot see them individually from a normal viewing distance. The colors we see on the screen are produced by selectively lighting up these three primary colors. Light from one third of the spectrum is combined with light from another third to produce other colors.

(a) What are the three additive primary colors? _____

(b) What dots would be lit to produce yellow on a TV set? _____

Answers: (a) red, green, and blue; (b) the green and red dots

5

An object appears a certain color because it absorbs some of the colors from the spectrum and reflects others. It subtracts out some of the colors of the spectrum and we see only what is left. In frames 3 and 4 we discussed forming new colors by adding lights together. Pigment colors combine differently, by a **subtractive process**. The primary colors that will allow us to obtain all other colors by mixing pigments are yellow, cyan, and magenta, as indicated in the second figure in frame 3. Each subtractive primary contains two-thirds of the spectrum. Yellow paint, for example, reflects the long-wavelength two-thirds of the spectrum and absorbs—or subtracts out—the short-wavelength third (the blue).

(a) What part(s) of the spectrum is/are absorbed by cyan paint? _____

(b) What part(s) is/are reflected? _____

Answers: (a) the red third;   (b) the blue and green thirds

6   Now consider what happens when we mix two pigments, say yellow and magenta paint. The yellow pigment absorbs the blue end of the spectrum, and the magenta absorbs the central portion, the green. Therefore, a third of the spectrum is still being reflected from our mixture of paint—the red third is reflected. Thus, a mixture of yellow and magenta paint yields red paint.

What color is produced by a mixture of cyan and yellow paint? _____

Answer: green (Yellow paint absorbs the blue third, reflecting the other two. Cyan paint absorbs the red third, reflecting the other two-thirds. Thus, between the two pigments, both the blue and red ends of the spectrum are absorbed. This leaves only green, the central portion, to be reflected.)

7 **UNDER COLORED LIGHTS**

You have seen that combining colored lights is an additive process, and that combining paint pigments to form different colors is a subtractive process. Suppose a yellow object is placed in red light. The object will appear red because a yellow object is capable of reflecting red light. (See part A of the second figure in frame 3.) This same object under blue light will appear black, however, because it cannot reflect blue light. Since no light is reflected, it appears black to us.

(a) Now consider the yellow object under cyan light. What color does it appear?

_____

(b) What color does a red object appear under magenta light? _____

(c) What color does a cyan object appear under magenta light? _____

Answers: (a) green (The object can reflect the green and red thirds of the spectrum. The cyan light shines the blue and green thirds onto it, but the blue is not reflected. The green is left.);   (b) red (The magenta light includes the red third.);   (c) blue (Blue is the only portion in the magenta light that can be reflected by the cyan object.)

8 WHY DON'T THINGS WORK THIS PERFECTLY IN PRACTICE?

In order to do the analysis we did in this chapter, we made some simplifying assumptions. If you go out and buy red paint and analyze its reflected light, you will find that its spectrum does not fit perfectly into the short-wavelength third of the visible spectrum. The paint may even reflect some light from the violet end. In general, actual pigments reflect a complicated spectrum, and one must know this spectrum well in order to predict accurately what color will appear when two pigments are mixed. If we understand the simple processes, however, we can better understand what happens in the more complicated real world.

(a) By what process do colors in lights combine? _____

(b) By what process do colors in pigments combine? _____

(c) By what process do colors in TV screens combine? _____

Answers: (a) additive; (b) subtractive; (c) additive

SELF-TEST

1. What is different concerning light waves that result in the phenomenon of color? _____

2. Name the additive primaries. _____

3. What color results when red and blue lights are shined on a white wall? _____

4. What color results from the addition of red and green light? _____

5. Name the subtractive primaries. _____

6. What color paint results from the mixing of magenta and cyan paints? _____

7. Are colors produced on a television screen by an additive or a subtractive process? _____

8. What color will a red object appear under yellow light? _____
   Under green light? _____   Under magenta light? _____

9. What color will a magenta object appear under green light? _____
   Under blue light? _____   Under yellow light? _____

10. Why do the examples of the additive and subtractive processes used in this chapter sometimes not work well in practice? _____

_____

_____

11. If cyan and red light are shined on a white wall, what color will the wall appear? _____   Magenta and green on the white wall? _____

ANSWERS

1. wavelength (or frequency) (frame 1)

2. red, green, and blue (frames 3, 4)

3. magenta (frames 3, 4)

4. yellow (frames 3, 4)

5. yellow, cyan, and magenta (frame 5)

6. blue (because it is common to both pigments) (frame 6)

7. additive (frame 4)

8. red; black; red (frame 7)

9. black; blue; red (frame 7)

10. Actual pigments normally do not reflect exactly one-third or two-thirds of the spectrum. They are more complicated than this, and so the results are more complicated. (frame 8)

11. white; white (frames 2, 3)

Appendix I
Scientific Notation:
Powers of Ten

Science deals with such large and small numbers that it is often inconvenient to write them in the standard form. The form used in science is called *scientific notation*, or powers of ten notation. Both large and small numbers can be expressed as a number between 1 and 10, times 10 raised to some power.

METHOD OF WRITING LARGE NUMBERS

Look at the following powers of ten.

$$10^1 = 10$$
$$10^2 = 100$$
$$10^3 = 1000$$
$$10^4 = 10,000$$
$$10^{23} = 1 \text{ followed by 23 zeros}$$

The number 2,553,000 is $2553 \cdot 10^3$, or in scientific notation, it is $2.553 \cdot 10^6$. To check that his latter number is the same as the original, move the decimal place six places to the right, as shown below.

2 5 5 3 0 0 0.

1, 2, 3, 4, 5, 6 places to the right

Here are a few more examples. You can check to make sure they are equal.

$$473,000 = 4.73 \cdot 10^5$$
$$760,000,000 = 7.6 \cdot 10^8$$
$$4.35 \cdot 10^3 = 4350$$
$$5.747 \cdot 10^2 = 574.7$$

METHOD OF WRITING SMALL NUMBERS

The power of ten tells you how far to move the decimal point to the right. The system works with negative powers also, but in this case the decimal point must be moved to the left. For example:

$$10^{-2} = 0.01$$

(We have taken the number 1.0 and moved the decimal point two places to the left.)

$$10^{-3} = 0.001$$
$$10^{-5} = 0.00001$$

More complicated numbers follow the same pattern.

$$4.6 \cdot 10^{-4} = 0.00046$$

The decimal point here has been moved four places to the left, the zeros being supplied. Compare these examples.

$$3.24 \cdot 10^{-5} = 0.0000324$$
$$0.00067 = 6.7 \cdot 10^{-4}$$
$$0.0000005 = 5 \cdot 10^{-7}$$

CALCULATIONS WITH SCIENTIFIC NOTATION (OPTIONAL)

To multiply and divide in scientific notation, follow two simple rules: (1) Handle the first part of the number and the power of ten part separately; and (2) multiply powers of ten by adding the powers, and divide by subtracting the powers. Some examples will illustrate.

$$(4 \cdot 10^2)(6 \cdot 10^3) = 24 \cdot 10^5 = 2.4 \cdot 10^6$$

Notice that although the interim answer was correct, it was changed to conform to standard scientific notation.

$$(1.3 \cdot 10^6)(2.2 \cdot 10^{-2}) = 2.86 \cdot 10^4$$

Here the 6 and the −2 were added, but since the second number was negative, this amounts to subtracting it from the first.

$$(1.33 \cdot 10^{-18})\,(3 \cdot 10^{5}) = 3.99 \cdot 10^{-13}$$
$$(1.33 \cdot 10^{-6})\,(3 \cdot 10^{4}) = 3.99 \cdot 10^{-2}$$
$$(1.33 \cdot 10^{-5})\,(3 \cdot 10^{-3}) = 3.99 \cdot 10^{-8}$$

Now for some division.

$$\frac{6 \cdot 10^{8}}{2 \cdot 10^{3}} = 3 \cdot 10^{5}$$

Here, 6 was divided by 2, and 10^{8} was divided by 10^{3} by subtracting the 3 from the 8.

$$\frac{2.6 \cdot 10^{9}}{2 \cdot 10^{-4}} = 1.3 \cdot 10^{13}$$

This results because 9 − (−4) = 9 + 4 = 13. Now a combination.

$$\frac{(2 \cdot 10^{5})(8 \cdot 10^{-2})}{4 \cdot 10^{3}} = \frac{16 \cdot 10^{3}}{4 \cdot 10^{3}} = 4$$

It doesn't matter whether the multiplication or the division is done first, but I chose to multiply first. You try it the other way for practice.

SELF-TEST

1. The brightest star in our sky is Sirius, which is 51,700,000,000,000 miles away. Write this number in scientific notation. _____

2. The speed of light is about $3 \cdot 10^{8}$ meters/second. Write this number in "standard" notation. _____

3. The mass of a proton is $1.67 \cdot 10^{-27}$ kilograms. Write this number without using powers of ten. _____

4. Write the number 0.000455 in scientific notation. _____

5. *Optional.* The mass of the proton is $1.67 \cdot 10^{-27}$ kilograms. What is the mass of a billion (10^{9}) protons? _____

6. *Optional.* The mass of the electron is $9.1 \cdot 10^{-31}$ kilograms. How many electrons are needed to have the same mass as one proton? (See the last question for the proton mass.) _____

7. *Optional.* The earth gets as far from the sun as $9.5 \cdot 10^{7}$ miles and Mars gets as close as $1.28 \cdot 10^{8}$ miles. How close does the earth get to Mars? (Hint: Subtract the two numbers.) _____

ANSWERS

1. $5.17 \cdot 10^{13}$ miles

2. 300,000,000 meters/second

3. 0.000,000,000,000,000,000,000,000,001,67 kilograms (The decimal has been moved 27 places, requiring 26 zeros between the decimal point and the first nonzero numeral.)

4. $4.55 \cdot 10^{-4}$

5. $1.67 \cdot 10^{-18}$ kilogram

6. $1.84 \cdot 10^{3}$ (or 1840)

7. $3.3 \cdot 10^{7}$ miles. Solution: The trick here is to write the numbers with the same power of ten. Using the power 7:

$$1.28 \cdot 10^{8} = 12.8 \cdot 10^{7}$$

Now subtract:

$$\begin{array}{r} 12.8 \cdot 10^{7} \\ -9.5 \cdot 10^{7} \\ \hline 3.3 \cdot 10^{7} \end{array}$$

Appendix II
The Metric System

In a few years, new books will not include an appendix on the metric system (though they may have one on the British system). The British system that we use will soon be a thing of the past for two reasons: First, it is desirable to have a single worldwide system, and second, the metric system is basically simpler than the British system.

LENGTH

The basic unit of length in SI units* is the meter. (A meter is 39.37 inches, a little longer than a yard.) The meter is divided into 100 centimeters, or into 1000 millimeters. Thus there are 10 millimeters in a centimeter. The only other unit of length which will be used in everyday language is the kilometer, which is 1000 meters. These lengths, as well as a number of other that are useful in science, are given below:

$$1 \text{ kilometer (km)} = 1000 \text{ meters (m)}$$
$$100 \text{ centimeters (cm)} = 1 \text{ m}$$
$$1000 \text{ millimeters (mm)} = 1 \text{ m}$$
$$10^6 \text{ micrometers (}\mu\text{m)} = 1 \text{ m}$$
$$10^9 \text{ nanometers (nm)} = 1 \text{ m}$$

Notice that all units are related to all others by a multiple of 10, and usually by a factor of 1000. This is why the metric system is basically simpler than the British system; there are no factors of 12, 3, or 5280.

---

*SI is the abbreviation for the French words for "International System." This is the metric system.

293

In the metric system, it is easy to change a measurement made in one unit of length to another unit of length. For example, 52 centimeters is 0.52 meters. (Notice that this is the same as saying that 52¢ = $0.52.) And 52 cm is 520 mm, or 520,000 µm, and so on.

MASS

In the British system we express weight in pounds. As explained in Chapter 2, there is a distinction between weight and mass, mass being the more fundamental quantity. And in the metric system we will speak of a person's mass rather than weight.

The basic SI unit of mass is the kilogram (kg). One kilogram weighs about 2.2 pounds on earth. Knowing what the prefix "kilo" means, you know that a kilogram is 1000 grams. Although we do not speak of centigrams, the units of milligram (1000 mg = 1 g) and microgram (10^6 µg = 1 g) are used in science.

Index

Absorption spectra, 224
Acceleration
    defined, 2–3
    equations, 4–5
    of gravity, 3–4, 17
    in Newton's laws, 16–21
    as a vector, 15–16
Air resistance, 19, 35, 36
Alternating current, 158–159
Ammeter, 171
Ampere, 151
Amplitude (of a wave), 106–107
Amplitude modulation, 192–193
AM radio, 191–192
Antenna, 190–191
Archimedes' principle, 73–75
Atmospheric pressure, 77–79
Atmospheric refraction, 243–244
Atom(s), 54–57
    Bohr model of, 55–57, 217–221
    electron configuration in, 55–57
    excited states of, 218, 221–223
    in a fluid, 72, 76
    in a solid, 64–65
Atomic mass, 60
Atomic number, 56
Atomic weight, 60

Avogadro's number, 60–61

Barometer, 79–80
Battery(ies), 151
    in series and parallel, 154–155
Beats (in sound), 127–128
Bimetallic strip, 87
Blackbody radiation, 213–214
Black light, 226
Bohr model of atom, 55–57, 217–221
Boiling, 96, 97
Bow wave, 110
Bright line spectrum, 221
British thermal unit (Btu), 91
Buoyancy, 73–75, 80

Calorie, 90
Camera, 261
Celsius temperature, 85–86
Centripetal force, 19–20
Change of state, 94–97
Charge, electric, 139–140
Charging
    by friction, 143
    by induction, 145–146
Circuit(s), 153
    series and parallel, 154–157

Circular motion, 19–20
Coherent light, 228
Collisions
    and conservation of momentum, 28–29, 37
    elastic and inelastic, 37
Color, 282–286
Compound, 52, 54
Compression
    in sound, 114–115
    in a spring, 108
Conduction, of heat, 97–99
Conductor, electrical, 144
Conservation
    of energy, 35–58
    of momentum, 26–31
Constructive interference, 126–128, 272,
    277–278
Continuous spectrum, 223–224
Convection of heat, 99
Conventional current, 150
Converging lens, 251–252
Coulomb, 139–140, 151
Coulomb's law, 139–140
Current
    alternating, 158–159
    direction of, 150
    induced, 178–180

Decibel, 117
Definite proportions, law of, 54, 55
Density, 65–66
Destructive interference, 126–128, 272,
    277–278
Diffraction
    of light, 171–172
    of sound, 124–125
Diffraction grating, 275–276
Diffuse reflection, 233
Diffusion, 76–77

Dispersion, 244–245
Diverging lens, 259–260
Domains, magnetic, 170
Doppler effect, 110–111, 134
Double slit interference, 272–274

Einstein, Albert, 215–217
Elasticity, 68–69
Electric charge, 139–140
Electric current
    alternating, 158–159
    direction of, 150
    induced, 178–180
Electric field, 141–143, 190
Electric meter, 171–172
Electric motor, 172–173
Electromagnet, 169
Electromagnetic spectrum, 194–195, 203
Electromagnetic theory, 205–206, 213–214
Electromagnetic wave, 191–195, 202–203
Electron, 55–56, 150
Electron volt, 219
Electroscope, 144–145
Element(s), 52, 53
    periodic table of, 57, 58
Ellipse, 44–45
Emission spectrum, 223–224
Energy, 32–35
    conservation of, 35–38
    electric, 151–152, 159–160
    gravitational potential, 32–34
    internal, 89–90
    kinetic, 34–35
    levels of (in atom), 218–221
    sound, 117–119
    thermal, 89–90
    in a wave, 109
Ether, "luminiferous," 206
Evaporation, 97

Excited states of atoms, 218, 221–223
Expansion, thermal, 87–89
Eye, 261–263
Eyepiece lens, 264

Fahrenheit temperature, 84–85
Farsightedness, 262
Field
    electric, 141–143
    magnetic, 165–170, 177–180, 191
Fluorescence, 225–226
Focal point and focal length, 251–252
Force
    centripetal, 19–20
    on current-carrying wire, 177–178
    net, 8
    and pressure, 67–68
Frequency
    of beats, 128
    of electromagnetic waves, 192–195
    fundamental, 130
    modulation, 193
    of pendulum, 105
    of sound, 116
    and wavelength, 106–108
Fusion, heat of, 94–95, 97

Galileo, 198
Gamma rays, 194
Gases, 76–78
    buoyancy in, 80
    diffusion through, 76–77
    molecules in, 76
    pressure of, 77–80
Gravity
    acceleration of, 3–4, 18
    law of, 42–43, 48–49
    and the moon, 18, 44

Ground, electrical, 150
Ground state of atoms, 222

Heat
    conduction of, 97–99
    convection of, 99
    expansion and, 87–89
    of fusion, 94–95, 97
    radiation of, 99–100
    specific, 91–92
    transfer of, 97–100
    of vaporization, 95–97
Hertz, 106
Hooke's law, 68–69
Huygens' principle, 200–201

Image(s)
    calculations of positions of, 257, 259, 268
    formation of, 254–256
    in a mirror, 233–234
    real, 255
    virtual, 255, 257–258
Impulse, 26
Incandescent light, 224
Index of refraction, 238–239
Induced current, 178–180
Induction
    charging by, 145–146
    of electric current, 178–181
    self-, 185–186
Inertia, 13–14
Infrared radiation, 99–100
Insulators
    electric, 144
    heat, 98
Intensity of sound, 116–119
Interference
    of light, 272–274, 277–278
    of sound, 125–128

Internal energy, 89–90
Ion, 56

Joule, 31, 152

Kelvin temperature, 86
Kepler's laws, 44–46
Kinetic energy, 34–35

Laser, 227–228
Length, units of, 293–294
Lens(es), 250–260
    of a camera, 261
    converging, 251–252
    diverging, 259–260
    of an eye, 261–263
    images formed by, 254–256
    of a microscope, 264
    objective and eyepiece, 264
    of a projector, 263–264
    of a telescope, 265–266
    vision correction with, 262–263
Lensmaker's formula, 253
Lenz's law, 181–182
Light
    diffraction of, 171–172
    dispersion of, 244–245
    emission by atoms, 218–221
    and the ether, 206
    interference of, 272–274, 277–278
    nature of, 209, 215–217
    polarization of, 278–280
    reflection of, 232–234
    refraction of, 234–240
    speed of, 198–200
Line spectrum, 221
Liquids, 72–76
    buoyancy in, 73–75
    diffusion through, 76–77

pressure in, 72–73
Longitudinal wave, 108–109
Loudness, 124
"Luminiferous" ether, 206

Magnetic field, 165–170, 177–180, 191
    of earth, 166
    produced by currents, 167–170
Magnetism, cause of, 170–171
Magnification, 256–257
    of microscopes, 264
    of telescopes, 265
Magnifier, 257–258
Mass, 14, 294
    vs. weight, 17
Maxwell, James Clerk, 202
Melting, 95
Mercury vapor lamp, 225
Meter, electric, 171–172
Metric system, 293–294
Michelson, Albert, 199–200, 207
Michelson–Morley experiment, 206–209
Microscope, 264
Mirage, 244
Mirror, 233–234
Mole, 60
Molecule(s), 54–55
    in a gas, 76
    in a liquid, 72, 96–97
    in a solid, 64–65
Momentum
    conservation of, 26–31
    definition of, 25
Moon
    acceleration of gravity on, 18
    gravity and the orbit of, 44
Motion
    circular, 19–20
    molecular, 89

projectile, 7–8
Motor, electric, 172–173

Nearsightedness, 262
Neutron, 55–56
Newton, Isaac
    and earth satellites, 47
    first law of, 13–14
    and Kepler's laws, 45–56
    and law of gravity, 42–44
    second law of, 14, 16–19
    and theory of light, 201–202
    third law of, 20–21
Newton (unit), 16
Nodes in a standing wave, 132

Objective lens, 264
Ohm, 153
Ohm's law, 152–153
Overtones, 130–131

Parallel circuit, 154–157
Pascal's principle, 75–76
Pauli exclusion principle, 57
Pendulum, 104
Period, 104–105
Periodic table of elements, 57–58
Phosphorescence, 227
Photoelectric effect, 203–206, 215–217
Photon, 216
Pitch (of sound), 129
Planck, Max, 215
Polarization of light, 278–280
Potential difference, 152
Potential energy, gravitational, 32–34
Power, 32
    electric, 159
    through transformers, 185

Powers of ten notation, 289–291
Pressure, 67–68
    atmospheric, 77–79
    in gases, 77–78
    in liquids, 72–76
    from solids, 67–68
    in a sound wave, 116–117
Primary colors, 284–285
Projectile motion, 7–8
Projector, 263–264
Proton, 55–56

Quality (of sound), 129–130
Quantum nature of light, 215–217

Radiation
    blackbody, 213–214
    of heat, 99–100
Radio, 191–192
Rainbow, 245–246
Rarefaction
    in sound, 114–115
    in a spring, 108
Ray of light, 236–237
Real image, 255
Reflection
    law of, 232
    of light, 232–234
    total internal, 241–243
Refraction
    atmospheric, 243–244
    index of, 238–239
    law of, 238–240
    of light, 234–240
    of sound, 120
Resistance
    of air, 19, 35, 36
    electrical, 153

Resistors, 153
    energy dissipation in, 159–160
    in series and parallel, 155–157
Resonance of sound, 119
Right-hand rule, 167, 177–178
Roemer, Olaus, 198

Satellites, 47
Scientific notation, 289–291
Self-induction, 185–186
Series circuit, 154–157
SI units, 293–294
Snell's law, 239
Solenoid, magnetic field of, 169–170
Solids, 64–65
    expansion of, 87–89
    molecules in, 64–65
Sonic boom, 134–135
Sound
    beats and, 128–129
    diffraction of, 124–125
    Doppler effect in, 134
    energy of, 117–119
    interference of, 125–127
    loudness of, 124
    quality of, 129–130
    refraction of, 120
    resonance in, 119
    speed of, 115
    as a wave, 114–115
Specific gravity, 67
Specific heat, 91–92
Spectroscopic analysis, 221
Spectrum
    absorption, 224
    continuous, 223–224
    electromagnetic, 194–195, 203
    emission (or line), 221
    visible, 282–283

Speed
    defined, 1–2
    of light, 198–200
    of sound, 115
    terminal, 19
Standing waves
    in a gas, 132–134
    in a spring, 131–132

Telescope, 265–266
Temperature
    and expansion, 87–89
    and heat, 89–90
    measurement of, 84–90
    and pressure, 77
Terminal speed, 19
Tesla, 168
Thermal coefficient of expansion, 88
Thermal conductivity, 98–99
Thermal energy, 89–91
Thermal expansion, 87–89
Thermometer, 87
Thin film interference, 277–278
Total internal reflection, 241–243
Transformer, 182–184
Transverse waves, 108–109
Triboelectric series, 143

Ultrasonic waves, 116
Units, metric, 293–294
Universal gravitational constant, 48

Vaporization, heat of, 95–97
Vectors, 5–11
    addition of, 8–11
Velocity, 5–7
    of a wave, 106, 107–108

Virtual image, 255–256, 257–260
Visible spectrum, 282–283
Vision correction with lenses, 262–263
Volt, 151–152
Voltage, 151–153
  in a transformer, 184–185

Watt, 32
Wave(s), 106–111
  electromagnetic, 191–195, 202–203
  in sound, 114–115
  standing, 131–134

Wavelength, 106–107
  of sound, 116
  in a spring, 106–107
Wave-particle controversy, 197–209
Weight, 17
  and the law of gravity, 43
Weightlessness, 47–48
Work, 31

X rays, 194

Young, Thomas, 202